AF764599

YOUTH TRANSITIONING FROM FOSTER CARE

ADRIENNE L. FERNANDES

Nova Science Publishers, Inc.
New York

For permission to use material from this book please contact us:
Telephone 631-231-7269; Fax 631-231-8175
Web Site: http://www.novapublishers.com

LIBRARY OF CONGRESS CATALOGING-IN-PUBLICATION DATA

Fernandes, Adrienne L.
Youth transitioning from foster care / Adrienne L. Fernandes.
p. cm.
ISBN 978-1-60692-069-5 (softcover)
1. Foster children--United States. 2. Foster home care--United States. I. Title.
HV875.55.F463 2008
362.73'30973--dc22

2008035975

Published by Nova Science Publishers, Inc. - New York

YOUTH TRANSITIONING FROM FOSTER CARE

CONTENTS

PREFACE

Nearly half of states have laws that explicitly permit the state child welfare system to continue providing foster care for children beyond the age of majority (usually no later than 19). However, the number of states that actually facilitate youth remaining in care beyond their 18th or 19th birthdays is significantly smaller. Over 20,000 young people have been emancipated from foster care annually from FY2002 through FY2006. While most young people have access to emotional and financial support systems throughout their early adult years, older youth in care and those who age out of care often face obstacles to developing independent living skills and building supports that ease the transition to adulthood. Older foster youth who return to their parents or guardians may continue to experience poor family dynamics or a lack of emotional and financial supports, and studies have shown that recently emancipated foster youth fare poorly relative to their counterparts in the general population on several outcome measures.

Recognizing the difficulties faced by older youth in care and youth emancipating from foster care, Congress created a new Independent Living initiative (P.L. 99-272) in 1986 to assist certain older foster youth as they enter adulthood. The legislation authorized mandatory funding to states under a new Section 477 of the Social Security Act. In 1999, the John H. Chafee Foster Care Independence Act (P.L. 106-169) replaced the Independent Living Program with the Chafee Foster Care Independence Program (CFCIP) and doubled the total annual funds available to states from $70 million to $140 million. The law also expanded the population of youth eligible to receive independent living services — with no lower age limit — and gave states greater flexibility in designing independent living programs. Independent living services can refer to assistance in obtaining a high school diploma, training in daily living skills, and training in financial management, among other services. Amendments to the CFCIP in FY2002 (P.L. 107-133) authorized discretionary funding for states to provide education and training vouchers to eligible youth.

Along with the CFCIP, federal child welfare law and other federal programs are intended to help older current youth in care and foster care alumni make the transition to adulthood. The federal foster care program has protections in place to ensure that older youth in care have a written case plan that addresses the programs and services that will assist in this transition, among other supports. Further, federal law authorizes funding for states to provide workforce assistance and housing to older foster youth. Despite these efforts and the resilience displayed by current and former foster youth, policymakers and child welfare practitioners have suggested that at a minimum, young people need better support to build stronger connections with caring adults before leaving foster care and should have the option to remain in care upon reaching their 18th or 19th birthdays.

Several bills have been introduced in the 110th Congress that propose greater assistance to older youth in foster care and those who have aged out, including legislation to provide federal foster care assistance for youth age 18 and older.

INTRODUCTION*

A significant number of youth spend at least some time in foster care during their teenage years. On the last day of FY2006, over 176,800 youth ages 13 to 17 comprised more than one third of the foster care caseload nationally.[1] Most teenagers who leave foster care are reunified with their parents or primary caretakers, adopted, or placed with relatives. Yet as many as 26,000 youth age out, or are emancipated, from care each year when they reach the age of majority in their states (usually no later than age 19).[2]

Youth who spend their teenage years in foster care and those who are likely to age out of care face challenges as they move to early adulthood. While in care, they may forego opportunities to develop strong support networks and independent living skills that their counterparts in the general population might more naturally acquire. Even older foster youth who return to their parents or guardians can still face obstacles, such as poor family dynamics or a lack of emotional and financial supports, that hinder their ability to achieve their goals as young adults. Perhaps the strongest evidence that youth who have spent at least some years in care during adolescence have not adequately made the transition to young adulthood is their poor outcomes across a number of domains. During their early adult years, these youth are much more likely than their peers to forego higher education, describe their general health as fair or poor, become homeless, and rely on public supports.

The federal government has recognized that older youth in care and those aging out are vulnerable to negative outcomes and may ultimately return to the care of the state as adults, either through the public welfare,

---

* Excerpted from CRS Report RL34499, dated May 21, 2008.

criminal justice, or other support systems. In 1986, Congress passed legislation to assist certain older youth in care under a new Independent Living program, enacted as part of P.L. 99-272. The legislation authorized mandatory funding to states under Section 477 of the Social Security Act and was made permanent in 1993 as part of P.L. 103-66. In 1999, the John H. Chafee Foster Care Independence Act (P.L. 106-169) replaced the Independent Living Program with the permanently authorized Chafee Foster Care Independence Program (CFCIP) and doubled the annual funds available to states from $70 million to $140 million. The law also expanded the population of youth eligible to receive independent living services, including youth who have left care through age 21, and gave states greater flexibility in designing independent living programs. Independent living programs are intended to assist youth prepare for adulthood, and may include assistance in obtaining a high school diploma, career exploration, training in daily living skills, training in budgeting and financial management skills, and preventive health activities, among other services. Further, the act required states to provide services to tribal youth on the same basis as other youth in the state. The act also encouraged youth in foster care to participate directly in designing their own activities that prepare them for independent living and further stated that youth “accept personal responsibility for living up to their part of the program.” Other related provisions in P.L. 109-133 enable states to use up to 30% of their CFCIP funds for housing on current and former foster youth ages 18 to 21, and to extend health care to these youth through a new Medicaid pathway known as the “Chafee option.” Amendments to the CFCIP in FY2002 (P.L. 107-133) authorized discretionary funding to states to provide vouchers for eligible youth through a new Education and Training Voucher (ETV) program.

Along with the CFCIP, federal child welfare law and other federal programs are intended to help current and former youth in foster care make the transition to adulthood. The federal foster care program has protections in place to ensure that older youth in care have a written case plan that addresses the programs and services they need in making the transition. Further, federal law authorizes funding for states and local jurisdictions to provide workforce support and housing to older foster youth and youth emancipating from care. Despite these efforts and the resilience of young people in care, policymakers and child welfare practitioners have suggested that at a minimum, young people should have the option to remain in foster care upon reaching their 18th birthdays and that they need more support to build genuine, permanent connections with caring adults before leaving care.

This report begins with a discussion of the characteristics of older foster youth in care and the types of outcomes experienced by youth who have recently emancipated. The report then provides an overview of the federal foster care system, including the Chafee Foster Care Independence program, and provisions in federal foster care law that are intended to help prepare youth for adulthood. The report goes on to discuss other federal support — through other programs — for youth aging out of care in the areas of education, health care, employment, and housing. The report seeks to understand how states vary in their approaches to serving older youth in care and those who are recently emancipated. A small number of states are known to extend foster care to youth ages 18 to 21 (and beyond in some cases) and less than half of states provide Medicaid coverage to former foster youth beyond age 18 through the Chafee pathway option. The report also intends to demonstrate that, despite negative outcomes for the group on average, many former foster youth are engaged in decisions about the services they receive and display resiliency. The report concludes with a discussion of issues that Congress may wish to consider, as well as pending legislation relevant to each of the issues.

Appendix A provides a summary of outcome statistics for youth who were in foster care, compared to youth in the general population; Appendix B summarizes state policies regarding youth remaining in care beyond age 18; Appendix C includes a description of foster care programs in selected states for youth ages 18 and older in foster care; Appendix D provides funding tables for the Chafee Foster Care Independence Program, including information about funds returned to the Federal Treasury; and Appendix E summarizes pending legislation that would amend the CFCIP.

Chapter 1

WHO ARE OLDER YOUTH IN FOSTER CARE AND YOUTH AGING OUT OF CARE?

Children and adolescents age 17 and younger can enter state child welfare systems due to abuse, neglect, or for some other reason, such as the death of a parent or child behavioral problems. Some children remain in their own homes and receive family support services, but many are placed in out-of-home settings, usually in a foster home, relative placement, or institution (e.g., residential treatment facility, maternity group home). This section provides a profile of older youth in foster care as well as outcomes experienced by young people who have spent some time in care as teenagers.

CHARACTERISTICS OF YOUTH IN FOSTER CARE

The Foster Care Dynamics report, a longitudinal study of children in 11 state child welfare systems from 2000 through 2005, provides detailed information about older youth who have been placed in foster care.[3] The study examined state administrative data to determine the typical trajectory of children across four age categories who first entered foster care during the five-year period: less than age one; one through five; six through12; and 13 through 17. The study found that teenagers make up a significant share of the foster care population; have shorter median lengths of stay relative to younger children; live in placements other than foster family homes; experience more placements in their first year in care than younger children; and most often exit care through reunification, although running away and reaching the age of majority are exit pathways for about 10% to 24% of these older youth, depending on their age.

From 2000 through 2005, about 350,000 children entered care in the 11 states, of whom 27% were teenagers ages 13 through 17. Youth in the age 13 through 17 category comprised the second largest share of children in care (n=94,965), only after children ages six through 12 (n= 95,676). The incidence rate for entering care for youth ages 13 through 17 was about 2.5 per 1,000 over the five-year period; this is compared to about 9.0 per 1,000 infants under age one; about 2.5 per 1,000 children ages one through five; and about 1.8 per 1,000 children ages six through 12. In three cohort years (2000-2001, 2002-2003, and 2004-2005), fifteen-year-olds comprised the second largest share of children in care by single-year age category (approximately 7%), only after infants under the age of one, who comprised 18% to 20% of the caseload.

Of those youth ages 13 through 17 who entered care from 2000 to 2005, 49% lived in congregate care (e.g, a residential school), 37% in foster family homes, and 13% in kinship care. The remaining two percent lived in an independent living arrangement or other arrangement. Across all age categories, 41% to 43% of children were moved within their first six months in foster care; however, a greater share of teenagers experienced multiple placements within their first six months of entering care. About 18% of youth ages 13 through 17 had two or more placements, compared with 12% to 16% of children in other age categories. Teenagers were also more likely to move to new living arrangements in the six to twelve months after entering care. While older youth in care had a shorter median length of stay than younger children in care, this median length of stay increased over the five-year period. The median lengths of stay for infants decreased from 18.7 months in 2000 to 17.6 months in 2004 (data were not available for 2005), and increased for youth ages 13 through 17 from 6.6 months in 2000 to 7.8 months in 2005.

Generally, teenagers were less likely to be adopted or placed with relatives as they got older and were more likely to run away and exit by reaching the age of majority or some other pathway such as independent living. With the exception of 17-year-olds, about the same share of teenagers were reunified as children ages three to 12. One-third to 46% of youth ages 13 through 17 exited to reunification. The balance of youth lived in an independent living arrangement or some other arrangement (12.4% to 15.1% for each age 13 through 17), ran away (6.7% to 11.7%), lived with a relative (5.1% to 8.4%), or reached the age of majority (0.9% to 23.9%). About two percent or less of the teenagers were adopted.

Although older youth tend to have shorter spells in foster care, they are more likely to re-enter care after their first exit. About 28% of youth who were discharged at ages 13 through 17 during 2000 to 2002 reentered care within one year, followed by infants, at 26%. Children ages one to five were least likely to re-enter care, at 15%.

OUTCOMES FOR YOUNG ADULTS FORMERLY IN FOSTER CARE

The transition to adulthood for all youth has becoming increasingly complex. During this period, young people cycle between attending school, working, and living independently. Many youth can rely on assistance from their families for financial and emotional supports during the transition. On average, parents give their children an estimated $38,000 — or about $2,200 a year — between the ages of 18 and 34 to supplement wages, pay for college tuition, and help with housing costs, among other types of financial assistance.[4] Parents also allow their adult children to live with them[5] and provide their children with non-material assistance, such as help with obtaining a driver's license, guidance on applying to college, advice on finances and establishing a new household, and connections to other caring adults in their communities. For older foster youth and those who have already aged out of care, this assistance is often not in place. Being in care may inherently cause instability because of multiple out-of-home placements, school transfers, and the challenge in maintaining relationships with parents and other kin. Youth may experience further instability if they cannot afford to live on their own or are unable to live with relatives or friends upon emancipation. In most states, youth may not remain in foster care beyond age 18, a time when most young people explore educational and career options.

Research on the transition for former foster youth is limited and most of the studies on outcomes for these youth face methodological challenges — they tend to be dated; include brief follow-up periods (e.g., no more than a year after exit from care); have low response rates, non-representative samples, and small sample sizes; and do not follow youth prior to exit from foster care.[6] Few studies include comparison groups to gauge how well these youth are transitioning to adulthood, in relation to their peers in the foster care population or general population. However, two studies — the Northwest Foster Care Alumni Study and the Midwest Evaluation of the

Adult Functioning of Former Foster Youth — have tracked outcomes for a sample of youth across several domains, either prospectively (following youth in care and as they age out) or retrospectively (examining current outcomes for young adults who were in care at least a few years ago) and comparing these outcomes to other groups of youth, either those who aged out and/or youth in the general population.

Both studies indicate that youth who spent time in foster care during their teenage years tended to have difficulty during the transition to adulthood and beyond. (The studies do not posit that foster care, per se, is associated with the challenges former foster youth face in adulthood. In fact, children tend to have a range of challenges upon entering care.[7]) The Midwest Evaluation has examined the extent to which outcomes in early adulthood are in fluenced by the individual characteristics of youth or their out-of-home care histories. The study found that still being in care, as opposed to having left care, and having certain other characteristics (i.e., having aspirations to graduate from college, feeling close to at least one family member, and expressing satisfaction with their experience in foster care) more than doubled the odds of working or being in school at age 19.[8]

The two reports exclude youth with developmental disabilities or severe mental illness that precluded them from participating, as well as youth who were incarcerated or in a psychiatric hospital.

Northwest Foster Care Alumni Study

Researchers with the Northwest Foster Care Alumni Study interviewed and reviewed the case files of 479 foster care youth who were in public or private foster care any time from 1988 to 1998 in Oregon or Washington.[9] On average, they interviewed youth who were 24.2 years old, with a range of 20 to 33 years old. The youth tended to be females (60% versus the 48% of females in foster care nationally in FY2006[10]); to have entered care as adolescents (11.1 years versus 8.1 years for children entering care in FY2006; nearly 60% of the youth in the study were age 12 and older at the time they entered care); and to have exited care between the ages of 15 and older (the mean age at exit was 18.5 years versus 9.8 years in FY2006 for foster youth nationwide).[11] Surveyed young adults experienced these outcomes even though most (83.6%) reported having access to "a lot" of child welfare services and supports, and about eight out of ten (81.5%) said that they felt loved while in care. These findings suggest that a confluence of factors, including the reasons they entered care, family dynamics, and access to services and supports before and after care, among many other variables, have likely influenced how well they function as adults.

The study compared the mental health status, educational attainment, and employment and finances for the foster care alumni to those of the general population.

- Mental health: Over 54% of foster care alumni had at least one mental health problem (depression, social phobia, panic disorder, and post-traumatic stress disorder, among others), compared to 22.1% of the general population.[12] About one quarter of the alumni experienced post-traumatic stress disorder (PTSD). This figure is greater than the prevalence of PTSD among Vietnam or Iraq war veterans — about 15%. Alumni tended to have similar recovery rates as their counterparts in the general population for major depression, panic syndrome, and alcohol dependency, but lower rates of recovery for other disorders such as generalized anxiety disorder, PTSD, social phobia, and bulimia.
- Education: While alumni have obtained a high school diploma or passed the general education development (GED) test at the same rates as 25-to-34-year-olds generally (84.5% versus 87.3%), they are much less likely to have a bachelor's degree — 1.8% versus 22.5% of all young people.
- Employment and finances: One third of alumni reported living below the poverty line, which is three times the national poverty rate. Further, almost 17% were dependent on Temporary Assistance for Needy Families (TANF), compared to 3% of the general population (although the high rate of participation in Oregon and Washington could have been due, in part, to TANF rules in those states). The alumni employment rate was 80%, while the general employment rate was 95%. Other indicators show that alumni were not financially secure. One third lacked health insurance (versus 18% of the general population) and 22% were homeless at least one day during the year after they left foster care (versus 1% of the general population who were homeless within the last year).

Midwest Evaluation on the Adult Functioning of Former Foster Youth

Few foster care alumni studies are prospective, meaning that they follow youth while in care through the time they leave care and beyond. The Midwest Evaluation is an ongoing study that tracks 600 or more (depending on the data collection wave) current and former foster youth in three states — Illinois, Iowa, and Wisconsin. All of the surveyed youth entered care prior to their 16th birthday.[13] Surveyed youth responded to researcher questions about outcomes in three data collection waves: at wave 1, when they were age 17 or 18, at which time most were in care; at wave 2 when they were ages 19 or 20, at which time some remained in care; and at wave 3, when they were ages 20 or 21 and no longer in care. Of those who remained in care beyond age 18, all were in Illinois, the only state of the three that retains court jurisdiction of foster youth (with the youth's permission) until age 21.[14] The Midwest Evaluation researchers expect to track youth outcomes at age 23 and possibly beyond.[15]

At wave 1, about half to 70% of youth in care reported received any one category of independent living services (educational services, employment/vocational support, budget and financial management support, housing services, health education services, and youth development services).[16] At wave 2, not more than half of the youth in care reported receiving at least one independent living service.[17] At wave 3, receipt of services decreased for youth, regardless of whether they remained in care. No more than approximately one third of the youth reported receiving at least one independent living service.[18]

For youth in the third wave, about one-third had left foster care within the previous 12 months, and another one-third had left care in the past three to four years; the balance of youth had left care one to three years prior or more than four years prior. Nearly nine out of ten were age 21 and 53.2% were female. Most (55.6%) identified as African American, followed by white (32.5%), multiracial (9.5%), and other races. Approximately 8% identified as Hispanic. More than two-thirds of youth (64.3%) reported feeling lucky to have been placed in foster care, compared with 9.7% who neither agreed nor disagreed with feeling lucky, and 26% who disagreed or strongly disagreed; about the same proportions of youth reported feeling satisfied with their experiences in foster care. Youth were also asked about receipt of independent living services across six domains: education, vocational training or employment, budgeting and financial management, health education, housing, and youth development such as conferences and leadership development activities. For each of the domains, about one-third of the youth or less reported having received relevant services since the wave 2 study and no more than one third reported having received relevant services since they were discharged from care. The researchers speculated that the services were available, but that the young adults did not perceive a need for the services or were unable to access them.

At ages 21 or 22, most youth reported strong family ties, with 94% having said that they felt somewhat or very close to at least one biological family member. The greatest shares of youth reported feeling very close to their siblings, followed by another relative (aunt, uncle, or cousin), grandparent, biological mother, and biological father. The surveyed youth were most likely to be in daily contact (in this order) with their siblings, biological mother, grandparents, and other relatives. Overall, more than half of all youth perceived that some or most of the time, they had social supports, such as someone to listen to him or her (66.1%), to help with favors (59.2%), to loan money (50.3%), and to encourage his or her goals (53.6%).

Emancipated Youth Compared to Youth in Care at Midwest Evaluation Wave 3

Findings from wave 3 of the Midwest Evaluation suggest that youth who remain in care as late as age 20 tend to experience an easier transition to adulthood than their counterparts who emancipate at age 18. The study found that the young adults from Illinois (who were more likely to remain in care than youth in Wisconsin and Iowa) were almost two times as likely to have ever attended college than their peers in the other two states; and were more likely to have higher earnings and delayed pregnancy. See the last section of this report for further information.

Youth at Midwest Evaluation Wave 3 Compared to Youth Age 21 Generally

For each of the three data collection waves, wherever possible, researchers asked the same questions that were taken directly from the National Longitudinal Survey of Adolescent Health ("Add Health"), a nationally representative survey that tracks a cohort of youth over time.[19] The former foster youth surveyed at wave 3 in the Midwest Evaluation were less likely to have attended college for at least one year compared to the Add Health youth (29.8% versus 52.9%). Youth in the Midwest evaluation who were not currently in school reported barriers to enrolling or staying in school, including that they lost interest (10.7%), became employed (12.0%), became a parent (12.0%), and other reasons. While youth formerly in care were almost as likely to report ever holding a job (95.1% versus 96.9%) as Add Health youth, a smaller share were currently employed (44.% versus 63.9%) and they had lower mean hourly wages ($8.85 versus $9.99) and mean annual incomes ($8,914 versus $12,728). About 52% of former foster youth reported having any savings or checking account, compared to 81% of youth generally. A greater share of former foster youth did not have enough money to pay rent (26.5%), compared to their counterparts in the Add Health survey (8.6%) and were more likely to report having received food stamps (50.2% versus 6.3% for females; 9.0% versus 0.0% for males).

Table A-1 in Appendix A presents the outcomes for youth surveyed in wave 3 and Add Health participants surveyed at age 21 across ten domains — living arrangements, educational attainment, employment and income, economic hardships, health, sexual behavior and pregnancy, relationships and family formation, involvement with the criminal justice system, transition to adulthood, orientation to the future, and mentoring. Table A-1 indicates that at age 21, former foster youth and youth generally shared some common characteristics, but that the former foster youth experience more negative educational and employment outcomes, among other outcomes.

Resiliency of Current and Former Foster Youth

Despite the generally negative findings from the two major evaluations on youth aging out of foster care, many youth have demonstrated resiliency by overcoming obstacles, such as limited family support and financial resources, and meeting their goals. As mentioned above, youth in the Northwest Evaluation have obtained a high school diploma or passed the general education development (GED) test at close to the same rates as 25-to-34-year-olds generally (84.5% versus 87.3%). Further, youth in the third wave of the Midwest Evaluation were just as likely as the general youth population to report being hopeful about their future. (See table A-1.) As described later in the report, these youth are active in making decisions about the independent living services they receive to help them prepare for adulthood.

Current and former foster youth are also working to make improvements to the child welfare system. At three hearings in the 110th Congress that have focused on older youth in foster care, these young people highlighted their struggles, successes, and their advocacy work on behalf of foster children. At a February 27, 2008 hearing before the Ways and Means' Subcommittee on Income Security and Family Support, a foster care alumni chronicled her life in care with eleven placements through her recent graduation from college. In her testimony, she urged Congress to improve conditions for children in foster care:

> I have accomplished a lot, but it is in spite of all of the uncertainty I experienced in foster care — not because of it. I want something better for the youth who are currently in the foster care system. I want them to have families to love and protect them and homes they know they can always return to. I want them to leave foster care to live with a family, a relative — someone who will be permanent in their lives. I do not want the youth currently in foster care to age-out of foster care with no family and no one to turn to for help or support.... Congress has the power to do something, and I ask you on behalf of all of my brothers and sisters who cannot be sitting here with me to do something now.[20]

The next chapter of this book describes the federal services and supports available to youth in foster care, and demonstrates how some states have varied in their approaches to serving older foster youth and youth who have recently emancipated from care.

Chapter 2

OVERVIEW OF FEDERAL SUPPORT FOR FOSTER YOUTH

Historically, states have been primarily responsible for providing child welfare services to families and children that need them. While in out-of-home foster care, the state child welfare agency, under the supervision of the court (and in consultation with the parents or primary caretakers in some cases), serves as the child's parent and makes decisions on his or her behalf that are to promote his or her safety, permanence, and well-being.

Safety refers to the state child welfare system's goal of ensuring that children in foster care are protected from further abuse or neglect. *Permanence* refers to the state's goal of ensuring that children do not spend too many of their formative years in a foster care placement, and that the state either returns them to their families quickly and safely or quickly finds another safe and permanent home for them. *Wellbeing* is inextricably linked to safety and permanency. The term refers to efforts by the child welfare system to promote positive outcomes for children in care, including education and physical and mental health outcomes, as well as supportive families. In most cases, the state relies on public and private entities and organizations to provide these services. The federal government plays a role in shaping state child welfare systems by providing funds and linking those funds to certain requirements.

Two programs under Title IV-E of the Social Security Act may provide support to older youth in foster care. Child welfare provisions under Title IV-E apply to children ages 18 and younger who are in state-supervised foster care, discussed below, and foster care youth who are expected to finish high school at age 19 (Section 472). The Chafee Foster Care Independence program, also discussed below, authorizes funding for states to provide independent living services to older foster youth and those who have emancipated from care (Section 477).

Chapter 3

FEDERAL FOSTER CARE PROGRAM

Federal support for foster care preceded, by several decades, the 1980 (P.L. 96-272) creation of the Title IV-E foster care program under the Social Security Act. However, the 1980 law established this support as an independent program to provide funding to states to support children in foster care. The law also stressed the importance of case planning and review to achieve permanence for children in care.

CASE PLANING AND REVIEW

Federal child welfare provisions under Title IV-B and Title IV-E of the Social Security Act require state child welfare agencies, as a condition of receiving funding under these titles, to provide certain case management services to all children in foster care. These include monthly case worker visits to each child in foster care (Section 422(b)(17)); a written case plan for each child in care (Section 475(1)); and procedures ensuring a case review is conducted not less often than every six months by a judge or an administrative review panel, and at least once every 12 months by a judge who must consider the child's permanency plan (Section 475(5)). Specific case plan and case review procedures pertain to older youth in care. For a child age 16 or older, the written case plan must include a description of the programs and services that will help the child prepare for the transition to independent living (Section 475(1)(D)), and the permanency plan hearing must consider "the services needed to assist the child to make the transition from foster care to independent living" (Section 475(5)(C)). The permanency hearing for all children, including those transitioning to independent living, must be conducted with the child in an age-appropriate manner (Section 475(5)(C)).

TITLE IV-E REIMBURSEMENT FOR FOSTER CARE

Title IV-E currently reimburses states for a part of the cost of providing foster care to eligible children and youth, who, because of abuse or neglect (or some other reason), cannot remain in their own homes and for whom a court has consequently given care and placement responsibility to the state. Under this program, a state may seek partial federal reimbursement to "cover the cost of (and the cost of providing) food, clothing, shelter, daily supervision, school supplies, a child's personal incidentals, liability insurance with respect to a child, and reasonable travel to the child's home for visitation" (Section 475(4)). States may also seek reimbursement for related costs of administration, child placement (e.g., case planning), training, and data collection. Under the Chafee Foster Care Independence Program (see below), states must certify that they will use Title IV-E foster care program funding (and Adoption Assistance program funding) to provide training relevant to foster parents and others (adoptive parents, workers in group homes, and case managers) to help them understand and address the issues confronting adolescents preparing for independent living and coordinating this training, where possible, with independent living programs.[21] Although case planning and review procedures (described above) apply to all foster children in state care, federal reimbursement to states under Title IV-E may be made only on behalf of a child who meets multiple federal eligibility criteria (Section 472), including those related to the child's removal and the income and assets of the child's family.

Eligible Placement Setting

For purposes of this report, the most significant eligibility criteria for the federal foster care program are the child's placement setting and age. Under the program, federal reimbursement of part of the costs of maintaining children in foster care may only be sought for children placed in foster family homes or child care institutions. States may not seek federal reimbursement of foster care costs for children who live in independent living arrangements[22] or who are in "detention facilities, forestry camps, training schools, or any other facility operated primarily for the detention of children who are determined to be delinquent."[23] For FY2006 (the most recent year data are available), almost 6,000 children (1% of the caseload) were placed in supervised independent living settings.[24]

Eligible Age

Further, once a child has reached his or her 18th birthday, he or she is no longer eligible for federal foster care assistance. The age limitation on Title IV-E eligibility is created by the program's eligibility link to the now-defunct Aid to Families with Dependent Children (AFDC) program.[25] Children qualified as dependents under the AFDC program until age 18. However, as was the case with AFDC, federal law does permit states to make continued claims for otherwise eligible foster youth until their 19th birthday provided that the youth is a full-time student and is expected to complete high school or an equivalent training program by age 19. States must have elected this option in its definition of "child" for purposes of the state's AFDC program.

A 2004 review of state laws by the American Bar Association found that just under half of states have laws that explicitly permit court jurisdiction to continue for foster children beyond the age of majority. Two states extend this jurisdiction to age 19, three states to age 20, and 17 states to age 21.[26] A survey of states conducted in 2006 by the National Child Welfare Resource Center for Youth Development also found that most states permitted youth to remain in care beyond their 18th birthdays, but only under certain circumstances (see below). However, the number of states that permit youth to remain in foster care beyond their 18th or 19th birthdays appears significantly larger than the number that actually facilitate youth remaining in care.[27]

National Child Welfare Resource Center for Youth Development Survey of Foster Care Beyond Age 18

In 2006, the National Child Welfare Resource Center for Youth Development at the University of Oklahoma (a contractor with the U.S. Department of Health and Human Services that provides training and technical assistance to states on foster care issues, including independent living) surveyed independent living coordinators in all states, Washington D.C., and Puerto Rico to determine whether emancipating foster youth are eligible to remain in foster care beyond age 18. The Resource Center received responses from 45 states and Washington D.C. However, the survey did not seek information about (1) the share of youth who actually remain in care beyond age 18; and whether (2) states encourage youth to remain in care; (3) financial supports are in place to support youth in care after their 18th birthday; (4) the extended time in care was authorized by state law; and (5) courts retained jurisdiction over the youth.

Table B-1 in Appendix B provides a list of the maximum age limit at which youth may remain in care and the conditions permitting youth to remain in care, as determined by the survey. Age 18 is the maximum age for youth to remain in care in one state (Florida). Six states (California, Nebraska, New Hampshire, Wisconsin, Vermont, and Utah) reported allowing youth to remain in care until age 19. Three states (Alaska, Iowa, and Michigan) permit youth to remain in care until age 20. Thirty-one states (Alabama, Arizona, Arkansas, Delaware, Georgia, Idaho, Illinois, Indiana, Kansas, Kentucky, Maine, Maryland, Minnesota, Missouri, Montana, New Mexico, New Jersey, Nevada, New York, North Carolina, North Dakota, Ohio, Oklahoma, Oregon, Pennsylvania, South Carolina, South Dakota, Virginia, Washington, West Virginia, Wyoming) and Washington, D.C. reported that they provide foster care custody through age 21 and two states (Massachusetts and Texas) permit foster care custody through age 22. Finally, two states (Colorado and Connecticut) permit foster care coverage through age 23. Hawaii, Louisiana, Mississippi, Rhode Island, Tennessee, and Puerto Rico did not respond to the survey. Four states that responded — Colorado, Connecticut, Nevada, and New York —indicated an age limit, but did not provide information on the conditions for remaining in care.

Sample of States with Known Programs That Provide Foster Care to Older Youth

Although over half of all states report that youth remain in foster care custody under certain circumstances until at least age 21, a much smaller number of states appear to encourage youth to do so. This is evidenced by the small number of youth ages 18 through 20 in foster care, as reported by HHS's Adoption and Foster Care Reporting and Analysis System

(AFCARS), the federal system that collects national foster care data. Data from AFCARS in FY2006 (the most recent year data are available) illustrates significant drop off between the number of youth age 17 (39,624; 8% of the foster care population), compared to youth age 18 (13,303; 3%); age 19 (5,488; 1%); and age 20 (3,316; 1%).[28]

The Congressional Research Service contacted four jurisdictions — Illinois, New York, Vermont, and Washington D.C. — that are known to retain youth in foster care at age 18 to 21 or 22.[29] These jurisdictions provide state foster care maintenance payments to fund foster care for older youth, and may use other funds besides Title IV-E maintenance payments. *Note that this is not an exhaustive review of states that provide the payments to youth beyond their 18th birthdays.* The four foster care programs for youth ages 18 and older vary. While Illinois, New York, and Washington, D.C. provide more traditional foster care maintenance payments to foster families on behalf of youth ages 18 to 21, Vermont provides stipends (much like foster care maintenance payments) to caring adults (including foster care families) who pledge to assist youth who have aged out of care.

In Washington, D.C., youth *must remain in care* until their 21st birthday unless they meet narrow criteria, whereas in Illinois and New York, eligible youth may decide to seek emancipation before reaching age 21. Former foster youth in Vermont are not required to participate in the program. Further, New York and Vermont require youth to be enrolled in an educational or workforce programs as a condition for remaining in care; Illinois and Washington D.C. do not. In all cases, with the exception of Vermont, youth ages 18 and older in care continue to be wards of the state. The juvenile courts retain jurisdiction and social workers make routine visits to assist youth in achieving their case goal, which is often independent living (another planned living arrangement). For additional information about the foster care programs for older youth in these states, see Appendix C.

Chapter 4

CHAFEE FOSTER CARE INDEPENDENCE PROGRAM

A second Title IV-E program, the Chafee Foster Care Independence program (CFCIP), provides primary federal support, including independent living services, to older youth in foster care and youth transitioning out of care.[30]

OVERVIEW

The John H. Chafee Foster Care Independence Act of 1999 (P.L. 106-169) replaced the prior law Independent Living Program with the Chafee Foster Care Independence Program. The 1999 law doubled the annual funds available to states for independent living services from $70 million to $140 million. To be eligible for funds, the act requires states to expand the population of youth who receive independent living services to include those who have "aged out" of foster care (until their 21st birthday) and those of any age in foster care who are expected to leave care without placement in a permanent family. Services may consist of educational assistance, vocational training, mentoring, preventive health activities, and counseling. States may dedicate as much as 30% of their program funding toward room and board for youth ages 18 to 21, including for those youth enrolled in an institution of higher education or who remain in foster care in states that provide care to youth until ages 19, 20, or 21.[31] Room and board are not defined in statute, but typically include food and shelter, and may include rental deposits, rent, utilities, and the cost of household startup purchases. CFCIP funds may not be used to acquire property to provide housing to current or former foster youth.[32] The act also required that youth in the program be actively involved in decisions about the services they receive.

In addition, the law changed the amount of earnings or other resources that foster youth may accumulate to assist in their transition to independent living. Eligibility for foster care maintenance payments under Title IV-E of the Social Security Act is based on whether the children's original families would qualify for AFDC, as it was in effect on July 16, 1996. Under those rules, children could not remain eligible for Title IV-E services if they accumulated assets of more than $1,000. P.L. 106-169, however, changed this asset limit to $10,000. The act also encouraged states to provide Medicaid coverage to youth ages 18, 19, and 20 who have emancipated from foster care by authorizing a new Medicaid eligibility pathway for "independent foster care adolescents," commonly called the "Chafee option" (see section on Medicaid in "Other Federal Support" below). In 2002 (P.L. 107-133), discretionary funds — up to $60 million annually — were authorized for eligible current and former foster care youth to receive education and training vouchers.

HISTORY

The CFCIP varies in its eligibility requirements and purpose from the former Independent Living Program, originally established in 1985 (P.L. 99-272). Congress created the former program under a new Section 477 of the Social Security Act to assist foster youth ages 16 to 18 who met the income and other eligibility criteria of the Title IV-E foster care program. The legislation authorized mandatory funding to states, and states were awarded a share of independent living funds based on the number of children receiving federal foster care payments in FY1984. In 1987 (P.L. 100-647) independent living eligibility was expanded to any foster care children age 16 or older (regardless of federal Title IV-E eligibility) and to certain youth for six months after leaving care. The Omnibus Budget Reconciliation Act of 1990 (P.L. 101-508) gave states *the option* of providing independent living services to current and former foster youth until age 21, and in 1993 (P.L. 103-66), Congress permanently authorized funding for the program at $70 million annually.

ELIGIBILITY FOR CFCIP BENEFITS AND SERVICES

The Chafee Foster Care Independence Act of 1999 required states to ensure that independent living programs serve children of "various ages and various stages of achieving independence" and use objective criteria for determining eligibility for benefits and services under the program. The act further specified that states are to provide services under the CFCIP for children who are "likely to remain in foster care until 18 years of age" or are "aging out of foster care." It also addressed the responsibilities of the states to consult with American Indian tribes and provide services to tribal youth.

Foster youth who are in runaway status or lose contact with their child welfare agency continue to be under the custody of the state, and therefore, are eligible for services upon returning. On the last day of FY2006, states reported to HHS that close to 12,200 (2%) children had run away from care. The 2006 data are similar to what states reported for the last days of FY2004 and FY2005.[33] Those youth in care who have been adjudicated through the juvenile justice system are eligible for CFCIP services as long as they are not in a detention or related facility.[34] (The juvenile corrections facility is responsible for all services for foster youth who are confined in a locked setting.) Finally, foster children who are not citizens may be eligible for CFCIP services while under state custody.

Youth Likely to Remain in Foster Care Until Age 18

Under the old Independent Living program, states could provide services to current foster youth ages 16 and 17 who were eligible for Title IV-E foster care maintenance payments, or to "other children in care," regardless of Title IV-E status. The Chafee Foster Care Independence Act removed reference to a minimum eligibility age and required states to provide supports to children "likely to remain in foster care" until age 18. This phrase is not defined in the act, and states are to create eligibility standards using objective criteria.

According to the National Foster Care Coalition, a child welfare advocacy organization, many states have developed indicators to help determine the likelihood that a child will remain in care until his or her 18th birthday.[35] For example, Louisiana determines a child's eligibility for independent living services by reviewing his or her case history, presenting problems, and individual case goals. Kentucky and Alaska require that services be provided concurrently with permanency planning for young people over the age of 14. Alaska further defines the level of such services that should be provided at age 14 and older.

States can provide services to any child age 17 and younger regardless of their placement in a kinship care home, family foster home, pre-adoptive home, or any other state-sanctioned placement so long as the child is in state custody. HHS's *Child Welfare Policy Manual* requires states that place children in foster care settings in other states to fund independent living services for foster youth ages 16 to 18 regardless of their placement in another state.[36]

Permanency Planning

The Adoption and Safe Families Act of 1997 (ASFA, P.L. 105-89) amended the Social Security Act at Section 475(5), which required that all young people in foster care have a permanency plan. ASFA established that, in addition to reunification, adoption, placement with a relative, and legal guardianship, "another planned permanent living arrangement" (APPLA) is specified as a permanency option, and may include independent living. While the CFCIP is intended to expand independent living services for older youth in foster care, Section 477 of the act encourages states to continue their efforts to achieve permanency for a young person, including by specifying that states should continue to locate and achieve placement in adoptive families for older youth in care. In his introductory remarks about the Senate version of P.L. 106-169 (S. 1327), Senator Jay Rockefeller described the intent of the legislation:

> "[A] youth's need for a family does not end at any particular age. Each of us can clearly recall times when we have had to turn to our own families for advice, comfort, or support long after our 18th or 21st birthdays. Many of us are still in the role of providing such support to our own children who are in their late teens or 20s. Therefore, an important provision in this Senate version of the Foster Care Independence Act states that Independent Living (IL) programs are not alternatives to permanency planning-young people of all ages need and deserve every possible effort made towards permanence, including adoption. It would be counterproductive to create any disincentive for adoption of teenagers."[37]

Youth Aging Out of Foster Care

Prior to the enactment of the CFCIP, states had the *option* to serve young people who had emancipated from care until age 21. The Chafee

Foster Care Independence Act *requires* states that receive CFCIP funds to provide independent living services to youth who have aged out of care between the ages of 18 through 21. According to HHS's *Child Welfare Policy Manual*, this requirement does not preclude states from providing services to other former foster care youth ages 18 to 21 who exited care prior to their eighteenth birthday.[38] Former foster youth continue to remain eligible for aftercare services until age 21 if they move to another state. The state in which the former foster youth resides — whether or not the youth was in foster care in that state — is responsible for providing independent living services to the eligible young person.[39]

Several states have developed recommitment policies for youth who have been discharged from care that specify a time limit in which they may be eligible for services. Under such a policy, youth who turn age 18 while in runaway status may re-enter care to receive services.[40] The Chafee Foster Care Independence Act further requires states to provide certain services to youth based on their age or whether they were in foster care at a specific age. States may provide room and board with CFCIP funds and Medicaid through the Chafee pathway only to those youth who are eligible for CFCIP services. Youth are eligible for an education and training voucher (until age 23) if they emancipate from care or were adopted from care at age 16 or older. However, to be eligible for a voucher at age 22 or 23, they must have received one at age 21.

American Indian Youth

The prior federal Independent Living Program did not specify that states consult with American Indian tribes or serve Indian youth in particular. P.L. 106-169 required that under the CFCIP, states must certify that each federally recognized Indian tribal organization in the state has been consulted about the state's independent living programs and that there have been efforts to coordinate the programs with these tribes.[41] The importance of tribal involvement was explained by Representative J.D. Hayworth during debate of the House version of P.L. 106-169 (H.R. 1802) in June 1999, when he said that tribes are in the best position to identify the needs of tribal youth and local resources available for these young people.[42]

Although tribal entities may not receive direct reimbursement from the federal government for foster care and related costs on behalf of Title IV-E eligible children, and must enter into intergovernmental agreements with states to receive this funding, tribes are not required to have entered into a Title IV-E agreement[43] with the state to participate in discussions with the state about its independent living programs and to access CFCIP funds and services.[44]

In addition to requiring states to consult and coordinate with Indian tribes, P.L. 106-169 also provides that the "benefits and services under the programs are to be made available to Indian children in the state on the same basis as to other children in the state." "On the same basis" has been interpreted by HHS to mean that the state will provide program services equitably to children in both state custody and tribal custody.[45]

National Child Welfare Resource Center for Youth Development survey of tribal consultation and independent living services

In 2007, the National Child Welfare Resource Center for Youth Development reviewed select parts of five-year state child welfare plans, known as Child and Family Service Plans (CFSP), to determine the extent to which (1) states have consulted tribes and involved tribal officials in the development of independent living programs and (2) tribal youth have accessed independent living services.[46] Based on a review of these five-year child welfare plans for 28 states, the NCWRCYD determined that the plans provided limited information about tribal consultation. In one-third of the plans, states did not provide a response about consulting tribes. (States may have provided this information in other parts of the plans that were not reviewed and/or may have in fact consulted widely with tribes.) However, some of the state plans detailed their consultation process with tribal entities. The NCWRCYD also reviewed annual child welfare progress reports (known as Annual Progress and Service Reports, which update the five-year plans) and found that 39 states reported consulting with tribes, but that only four states elaborated with additional information.

THE ROLE OF YOUTH PARTICIPANTS

Section 477 of the Social Security Act requires that states ensure youth in independent living programs participate directly in designing their own program activities that prepare them for independent living and further that they "accept personal responsibility for living up to their part of the program." This language builds on the positive youth development approach to serving youth.[47] Youth advocates that support this approach view youth as assets and promote the idea that youth should be engaged in decisions about their lives and communities.

States have also taken various approaches to involving young people in decisions about the services they receive. These include annual conferences, with young people involved in conference planning and participation; youth speakers' bureaus, with young people trained and skilled in public speaking; youth or alumni assisting in the recruitment of foster and adoptive parents; and young people serving as mentors for children and youth in foster care, among other activities.[48] Some states have also established formal youth advisory boards to provide a forum for youth to become involved in issues facing youth in care and aging out of care.[49] Youth-serving organizations for current and former foster youth, such as Foster Club, provide an outlet for young people to become involved in the larger foster care community and advocate for other children in care. States are not required to utilize life skills assessments or personal responsibility contracts with youth to comply with the youth participation requirement, although some states use these tools to assist youth make the transition to adulthood.[50]

ADMINISTRATION OF THE PROGRAM

States administer their independent living programs in a few ways. Some programs are overseen by the state independent living office, which employs an independent living coordinator and other staff. For example, in Maine, the state's independent living manager oversees six specialized life skills education coordinators assigned to cover all of the state's district offices for the Department of Health and Human Services. In some states, like California, each county (or other jurisdiction) administers its own program with some oversight and support from a statewide program. Other states, including Florida, use contracted service providers to administer their programs. Many jurisdictions have partnered with private organizations to help fund and sometimes administer some aspect of their independent living programs. For example, the Jim Casey Youth Opportunities Initiative has provided funding and technical assistance to ten cities to provide financial support and training to youth exiting care.[51]

Education and Training Vouchers

As mentioned above, in 2002, Congress passed legislation (P.L. 107-133) to authorize discretionary funding to eligible current and former foster care youth for education and training vouchers worth up to $5,000 annually per youth (states may determine the annual period to which to apply the $5,000 ceiling[52]). Youth qualify for a voucher if they are eligible for CFCIP services or were adopted from foster care after 16 years of age. Youth eligible for CFCIP services includes those who are ages 18 to 21 who have left foster care because they have aged out; youth likely to remain in foster care until age 18, as determined by the state; and former foster care recipients age 21 and younger, as determined by the state.

The vouchers are available for the cost of full-time or part-time attendance at an institution of higher education, as defined by the Higher Education Act of 1965.[53] (ETV funds may be used to pay for health insurance, which may cover mental health services; but they may not be used to pay directly for a doctor's visit or dental insurance.[54]) A current fiscal year's ETV funds may not be used to finance a youth's educational or vocational loans incurred prior to that current fiscal year.[55] Students may receive the vouchers if they are in good standing and making progress toward completing their program or graduating, though states may have additional requirements such as periodically meeting with a caseworker or limiting the funding to a certain number of semesters.[56] Only youth receiving a voucher at age 21 may continue to participate in the voucher program until age 23.

Funding received through the ETV program does not count toward the student's expected family contribution, which is used by the federal government to determine a student's need for federal financial aid (even for those students who are classified as *independent*, meaning that their parental financial information is not included in the financial aid analysis). However, the total amount of education assistance provided under the CFCIP and other federal programs may not exceed the total cost of attendance, and students cannot claim the same education expenses under multiple federal programs.

Administration of the Program

States and counties may use ETV dollars to fund the vouchers and the costs associated with administering the program, including for salaries, expenses, and training of staff who administer the state's voucher program.

States are not permitted to use Title IV-E Foster Care or Adoption Assistance program funds for administering the ETV program.[57] They may, however, spend additional funds from state sources or other sources to supplement the ETV program or use ETV funds to expand existing post-secondary funding programs.[58] Several states have scholarship programs, tuition waivers, and grants for current and former foster youth that are funded through other sources.[59]

Generally, states administer their ETV program through their independent living program. Some states, however, administer the program through their financial aid office (e.g., California Student Aid Commission) or at the local level (e.g., Florida, where all child welfare programs are administered through community-based agencies). Some states contract with a non-profit service provider, such as the Orphan Foundation of America or the Student Assistance Foundation. For example, the Orphan Foundation of America (OFA), a non-profit child welfare organization, contracts with nine states (Alabama, Arkansas, Colorado, Indiana, Maryland, Missouri, North Carolina, Ohio, and New York) to administer their ETV programs. OFA's administrative fee is capped at 12% of funds disbursed on behalf of a state. OFA is involved in all aspects of program administration, including identifying eligible youth for the program; maintaining a database with financial records, youth demographics, and youth outcomes for the nine states; training child welfare staff about the ETV program; and providing over-the-phone financial aid counseling to ETV recipients in the nine states. Students identified at risk of dropping out because of their grades receive academic counseling by OFA staff.

Youth Enrolled in the Program

Youth may enroll in the program if they are in care or have aged out of care. The state with the placement and responsibility for a youth in foster care is to provide the voucher to that youth. The state must also provide a voucher to any youth who is currently receiving a voucher and moves to another state for the *sole* purpose of attending an institution of higher education.[60] If a youth permanently moves to another state after leaving care and subsequently enrolls in a qualified institution of higher education, he or she can apply for a voucher in his or her current state of residence.[61]

HHS's Administration for Children and Families reported that states provided vouchers to 11,365 youth in FY2006 and to 12,692 youth in FY2007 under the Education and Training Voucher program.[62] In the nine states for which the Orphan Foundation of America administers ETV funds, it collects information on the type of schools ETV recipients attend, their demographics, and their outcomes. In FY2007, over 4,260 youth were eligible for the vouchers in those nine states, and just over half of these youth — 2,400 (56.7%) — received a voucher. More than half (55.4%) of the youth attend universities; 35.8% attend community colleges; and 8.7% attend technical and speciality schools. (Many students pursue technical training through community college programs). Based on demographic data collected of the ETV recipients in the nine states, 58% were African American, 28% were white, 7% were Latino, 5% were of two or more races, 1% were Native American, 1% were Asian American, and less than 1% (11 recipients) were Pacific Islander. The majority of youth recipients were female (68%) and just over half (54.5%) of the youth were ages 19 or 20. Nearly 18% of the youth were parents; the average age at which the youth became a parent was 17 years and nine months.

FUNDING FOR STATES

States must provide a 20% match (in-kind or cash) to receive their full federal CFCIP and ETV allotment. CFCIP funds are often mixed with state, local, and other funding sources to provide a system of support for youth likely to age out of care and those who have emancipated. To be eligible for CFCIP general and ETV funds, a state must submit a five-year plan (as part of its Child and Family Service Plan (CFSP) and Annual Progress and Service Report (APSR)) to HHS that describes how it intends to carry out its independent living program. The box below lists the certifications that the state must make when submitting its plan. Under Section 477 of the Social Security Act, HHS must approve a state's plan for independent living services if it contains all of the required materials. The plan must be submitted on or before June 30 of the calendar year in which the plan is to begin. States may make amendments to the plan and notify HHS within 30 days of modifying the plan. HHS is to make the plans available to the public. Mandatory and discretionary funds provided under the CFCIP may only supplement, rather than supplant, any funds from other federal sources (e.g., Social Services Block Grant or Runaway and Homeless Youth program) or non-federal sources that may be available for independent living programs in the state.

Use of Funds

States may apply to receive mandatory funds for the six purposes specified in the CFCIP:

(1) to identify children who are likely to remain in foster care until 18 years of age and to help these children make the transition to self-sufficiency by providing independent living and related services;

(2) to help children who are likely to remain in foster care until 18 years of age receive the education, training, and services necessary to obtain employment;

(3) to help children who are likely to remain in foster care until 18 years of age prepare for and enter postsecondary training and education institutions;

(4) to provide personal and emotional support to children aging out of foster care, through mentoring and positive interactions with caring adults;

(5) to provide financial, housing, counseling, employment, education, and other appropriate support and services to former foster care recipients between ages 18 and 21 years to complement their own efforts to achieve self-sufficiency and to assure that program participants recognize and accept personal responsibility for making the transition from adolescence to adulthood; and

(6) to make available education and training vouchers for youth who have aged out of care.

As described in HHS's *Child Welfare Policy Manual*, states may also use CFCIP funding to establish trust funds for youth eligible under the program (Section 3.3E, Question 1). However, states may not use CFCIP funds to train foster parents, workers in group homes, and case managers in issues confronting adolescents. States may use funding under the Title IV-E Foster Care program or Adoption Assistance program for this purpose.

CFCIP and ETV funds are distributed to each state based on its proportion of the nation's children in foster care. The Consolidated Appropriations Act of 2008 (P.L. 110-161) provides $140 million in mandatory funding for the CFCIP and $45.4 million for the ETV program. Table D-1 in the Appendix provides the CFCIP and voucher allotments for each state in FY2007 and FY2008.

CFCIP State Plan Requirements

To receive funds under the CFCIP, a state must describe in its CFCIP plan how it will
— design and deliver programs to achieve the program purposes;
— ensure statewide, although not necessarily uniform, coverage by the program;
— ensure that the programs serve children of various ages and at various stages of achieving independence;
— involve the public and private sectors in helping adolescents in foster care achieve independence;
— use objective criteria for determining eligibility for and ensuring fair and equitable treatment of benefit recipients;
— cooperate in national evaluations of the effects of the programs in achieving the purpose of the CFCIP.
The state must also certify that it will
— provide assistance and services to eligible former foster youth;
— use room and board payments only for youth ages 18 to 21;
— expend not more than 30% of CFCIP funds on room and board for youth ages 18 to 21;
— use funding under the Title IV-E Foster Care program and Adoption Assistance program (but not the CFCIP) to provide training to help foster parents and others understand and address the issues confronting adolescents preparing for independent living and coordinate this training, where possible, with independent living programs;
— consult widely with public and private organizations in developing the plans and give the public at least 30 days to comment on the plan;
— make every effort to coordinate independent living programs with other youth programs at the local, state, and federal levels, including independent living projects funded under the Juvenile Justice and Delinquency Prevention Act, abstinence education programs, local housing programs, programs for disabled youth, and school-to-work programs offered by high schools or local workforce agencies;
— consult each Indian tribe about the programs to be carried out under the plan, that there have been efforts to coordinate the programs with such tribes, and that benefits and services under the programs will be made available to Indian children in the state on the same basis as other children in the state;
— ensure that eligible youth participate directly in designing their own program activities that prepare them for independent living and that they accept personal responsibility for living up to their part of the program;
— establish and enforce standards and procedures to prevent fraud and abuse in the programs carried out under its plan;
— ensure that the ETV program complies with the federal program requirements, including that (1) the total amount of education assistance to a youth provided through the ETV program and under other federal and federally supported programs does not exceed the total cost of attendance and (2) does not duplicate benefits under the CFCIP or other federal or federally-assisted benefit program.

Source: Section 473 of the Social Security Act.

Hold Harmless Provision

Section 477 of the Social Security Act's includes a "hold harmless" clause that precludes any state from receiving less than the amount of general independent living funds it received under the former independent living program in FY1998 or $500,000, whichever is greater. (There is no hold harmless provision for ETV funds.) The general funding for independent living services doubled nationally with the implementation of the CFCIP, however, the percentage change in funds received varies across states. This is because the distribution of funding was changed to reflect the most current state share of the national caseload (instead of their share of the 1984 caseload in all previous years). Some states receive nearly the same level of funding while others receive much more under the CFCIP (see table D-2 in the Appendix).[63] In FY2007 (the latest year for which final data are available), three states (the District of Columbia, Louisiana, and New York) received the same funding amount they received in FY1998. Eight states (Alaska, Delaware, Idaho, New Hampshire, North Dakota, South Dakota, Vermont, and Wyoming) each received $500,000 because their FY1998 allotments were this amount or less.

Unused Funds

States have two years to spend their CFCIP and voucher funds. For instance, funds allotted for FY2008 may be spent in FY2008 or FY2009. If a state does not apply for all of its allotment, the remaining funds may be redistributed among states that needs these funds as determined by HHS. If a state applies for all of its CFCIP allotted funds but does not spend them within the two-year time frame, the unused funds revert to the federal treasury.

Table D-3 in Appendix D shows the FY2005 (the latest year for which data for returned funds are available) final allocations, dollars expended, and dollars returned to the federal treasury for general CFCIP allotments. That fiscal year, the 50 states, Puerto Rico, and Washington D.C. were allocated a combined total of $137.9 million in general CFCIP funds, about one percent of which was returned to the treasury. No states returned all of their funds.

Thirteen states (Delaware, Indiana, Kentucky, Maryland, Massachusetts, Minnesota, Mississippi, Missouri, North Carolina, North Dakota, Texas, Virginia, and Wyoming) each returned between 0.1% and 26.4% of their allocations.

States initially received funds for education and training vouchers in FY2003. Table D-4, also in Appendix D, shows that in FY2005 (the latest year for which data are available), the 50 states, Puerto Rico, and Washington D.C. received $45.9 million in funds for the vouchers, of which 26 states and Puerto Rico collectively returned 11.6% of those funds. Final data are not available for Alaska and Puerto Rico, and therefore the total share of returned funds is not final. States did not report to HHS the reasons for not spending ETV funds. Based on Congressional Research Service (CRS) discussions with HHS staff, OFA staff, and a small number of states, the reasons may include (1) the lack of infrastructure to absorb and distribute funds, including training child welfare workers and knowledge about educational financial aid; (2) mechanisms for distributing funds to youth and tracking youth outcomes; (3) federal fiscal year deadlines; and (4) youth's knowledge about the ETV program.[64]

National Youth in Transition Database

Section 477 requires that HHS consult with state and local public officials responsible for administering independent living and other child welfare programs, child welfare advocates, Members of Congress, youth service providers and researchers to (1) "develop outcome measures (including measures of educational attainment, high school diploma, avoidance of dependency, homelessness, non-marital childbirth, incarceration, and high-risk behaviors) that can be used to assess the performance of states in operating independent living programs;" (2) identify the data needed to track the number and characteristics of children receiving services, the type and quantity of services provided, and state performance on the measures; and (3) develop and implement a plan to collect this information beginning with the second fiscal year after the passage of P.L. 106-169. The law further required that by December 2000, HHS was to submit to the House Ways and Means Committee and Senate Finance Committee a report detailing a plan and timetable for collecting the data from the states and a proposal to impose penalties of not less than 1% or more than 5% of the states' CFCIP funding for states that do not comply with the data collection requirements, based on the degree of noncompliance.

In its September 2001 *Report to the Congress*, "Developing a System of Program Accountability Under the John H. Chafee Foster Care Independence Program," HHS outlined a plan and timetable for developing a data collection system. The report stated that HHS administrators and an

HHS working group consulted with stakeholders and reviewed child welfare literature to identify the data elements to be collected and reported in the database, as well as the outcomes relevant to assessing the CFCIP. The report discussed a penalty structure for states not in compliance with the reporting requirements. HHS also established a pilot test in 2001 involving seven states and an Indian tribe. In each of the jurisdictions, caseworkers collected data about older youth, identified unclear data elements, and described any difficulties encountered while collecting the information. According to HHS, the pilot test enabled the agency to assess the burden for workers collecting the data and to learn if the capacity to report data varied significantly across states and agencies.

Acting on the advice of the Department's General Counsel, HHS decided to produce formal regulations for this mandatory data collection process instead of following their initial plan of producing Program Instructions. The September 2001 report stated that HHS would publish the final outcomes and measures in the *Federal Register* and seek public comment by May 2002. The agency anticipated administering the database nationally in October 2003. However, a notice of proposed rulemaking (NPRM) for the data collection system was not published until July 14, 2006.

The NPRM provided for a 60-day comment period during which HHS heard from states and child welfare advocates. After this period, the final regulations underwent a full agency and department review as well as Office of Management and Budget (OMB) clearance before final publication on February 26, 2008.[65] The final rule establishing the NYTD became effective April 28, 2008, sixty days after publication and requires states to report data on youth beginning in FY2011.

HHS will use the NYTD to engage in two data collection and reporting activities.[66] First, states will collect information twice each fiscal year on eligible youth who currently receive independent living services whether they continue to remain in foster care, were in foster care in another state, or received child welfare services through an Indian tribe or privately operated foster care program. These youth are known as *served youth*. Second, states will also collect information on foster youth on or about their 17th birthday, two years later on or about their 19th birthday, and again on or about their 21st birthday. Foster youth age 17 are known as the *baseline youth* and at ages 19 and 21 are known as the *follow-up youth*. These current and former foster youth will be tracked regardless of whether they receive independent living services at ages 17, 19, and 21. States have the option of tracking a sample of youth who participated in the outcomes collection at age 17 to reduce the data collection burden. Information will be collected on a new group of foster youth age 17 every three years.

Consistent with the statutory requirement developed by Congress in P.L. 106-169, HHS proposes to penalize any state not meeting the data collection procedures for the NYTD from 1% to 5% of its annual Chafee fund allotment, which includes any allotted or re-allotted funds for the general CFCIP program only. The penalty amount will be withheld from a current fiscal year award of the funds. ACF will evaluate a state's data file against data compliance standards, provided by statute. However, states will have the opportunity to submit corrected data.

EVALUATION OF INNOVATIVE CFCIPS

Section 477 provides that HHS is to conduct evaluations of independent living programs funded by the CFCIP deemed to be innovative or of national significance. The law reserves 1.5% ($2.1 million) of total CFCIP funding annually for these evaluations, as well as CFCIP-related technical assistance, performance measurement, and data collection. For FY2007, HHS spent $1.4 million on research and evaluation and the remaining was set aside for technical assistance and other program support.

HHS has contracted with the Urban Institute and its partners to conduct the five-year Multi-Site Evaluation of Foster Youth Programs. The goal of the evaluation is to determine the effects of independent living programs funded by P.L. 106-169 in achieving key outcomes, including increased educational attainment, higher employment rates and stability, greater interpersonal and relationship skills, reduced non-marital pregnancy and births, and reduced delinquency and crime rates.

Researchers initially conducted an assessment to identify programs that could be evaluated rigorously (i.e., random assignment to treatment and control groups), as required under Section 477 of the Social Security Act. The research team determined that it could use random assignment at four innovative programs in California and Massachusetts — an employment services program in Kern County, California; a one-on-one intensive, individualized life skills program in Massachusetts; and, a tutoring/mentoring program and a classroom-based life skills training program, both in Los Angeles County, California.[67]

According to HHS, the researchers have completed collecting data from the Los Angeles sites and continue to collect data for the other two sites.[68] The data collection process will be complete in Kern County in 2008 and in

Massachusetts in 2009. Findings from the two Los Angeles sites are expected to be released in 2008; findings from the Kern County site are expected to be released in 2009; and findings from the Massachusetts site are expected to released in 2010. A final study will be released in 2010.

The 1,400 youth participating in the evaluation were assigned to intervention and control groups, and have been surveyed at three points: baseline, one year after baseline, and two years after baseline. The researchers have conducted in-person interviews with the youth to obtain information on youth characteristics, program interventions and services, and intermediate and longer-term outcomes. Researchers have also conducted interviews with program administrators, community advocates, and directors of community provider agencies. Further, the research team has held focus groups with youth, independent living program staff, and other agency staff responsible for referring youth to the programs. The team is using extracts of state administrative data to determine child and family demographics, child welfare placement history, physical and mental health status, and delinquency history. Data specific to each site (such as information from tutor contact logs at the Los Angeles tutoring/mentoring program and length of outreach services and service referrals at the life skills program in Massachusetts) is also being collected by the team.

PART Review

In calendar year 2004, the CFCIP was reviewed through the U.S. Office of Management and Budget's Program Assessment Rating Tool (PART) process.[69] The evaluation concluded that program results were "not demonstrated" because the CFCIP lacked long-term performance measures and time frames for these measures, as well as adequate progress in achieving its annual performance goals; and because some states did not use all of their CFCIP or ETV funding. OMB made the determination about the performance measurement because the National Youth in Transition Database, which will provide long-term performance and annual performance measurement beginning in FY2011, had not yet been implemented. The PART review also found that no independent evaluations of the program have been routinely conducted. While the Multi-Site Evaluation will provide information about outcomes for youth who receive independent living services, according to the PART review, it is not sufficiently national in scope.

The PART has established two goals for the program:

- Promote the efficient use of CFCIP funds by 1) increasing the percentage of states that completely expend their allocations within the two-year expenditure period, and 2) decreasing the percentage of funds that remain unexpended by states within that period. For FY2007, the targets for these measurements are a 7% increase in the number of states (over the previous year) that completely spent CFCIP allocations; and a 20% decrease (from the previous year) in the amount of funds remaining unexpended. The actual figures are expected to be reported in January 2009.
- Increase the percentages of CFCIP youth who avoid high-risk behaviors which might otherwise lead to criminal investigations and incarceration. The baseline measurement will be taken in FY2011, the first year the NYTD is implemented.

Assessments of the CFCIP

There does not appear to be a clear picture of the types of services provided through CFCIP-funded programs, although the National Youth in Transition Database and the evaluation of innovative independent living programs will provide a national overview about the number of youth served, the types of services provided, and best practices in assisting current and former foster youth make the transition to adulthood. Three sources — an evaluation of independent living services described in states' Child and Family Services Review documents, a GAO report on the CFCIP, and a report on ETV programs in six states — provide some insight into how states carry out their independent living programs.[70]

Child and Family Services Review

The U.S. Department of Health and Human Services determines state compliance with federal child welfare policies, and helps to ensure that positive outcomes are achieved for children and families involved in the child welfare system, by conducting reviews of state and child welfare programs.

The most comprehensive component of HHS's review system is the Child and Family Services Reviews (CFSR).[71] Conducted by the Children's Bureau, the reviews assess state conformity with certain requirements of Title IV-B and Title IV-E.[72] The first round of reviews was conducted between 2001 and 2004 in all 50 states, the District of Columbia, and Puerto Rico, and a final report was prepared discussing the findings in each jurisdiction.[73] This initial round found that no state's child welfare programs met the criteria that HHS established as demonstrating "substantial conformity" with all of federal child welfare policy requirements. As a result, all states have or are implementing Program Improvement Plans (PIPs). To avoid financial penalties associated with noncompliance, states must meet the improvement goals established in their PIP. The second round of reviews for some jurisdictions are underway.

To achieve substantial conformity with federal child welfare policy, states must achieve seven outcomes related to the safety, permanency, and well-being of children and they must demonstrate they have in place child welfare systems to achieve these goals. A review team composed of federal and state evaluators uses 45 items — or performance indicators — to guide the team through an evaluation of the state's performance. Though none of the performance indicators specifically target older children in care, the review team assess how well states meet the needs of all children in foster care, including adolescents, through multiple indicators.[74]

The CFSR second round review began in 2007 and requires federal and state evaluators to interview, at the state-level, youth being served by the state child welfare agency, especially those youth who are eligible to receive independent living services; HHS also recommends that evaluators interview the state youth service agency.[75] Further, at the local level, the evaluators are to interview youth being served by the local child welfare agency, particularly those eligible for independent living services.

Analysis of first round CFSR findings

An analysis of the CFSR findings for 45 jurisdictions found strengths in a few states related to increased involvement of youth in case planning and other relevant activities, as well as an increased focus on permanency issues for adolescents in care, including promoting family connections for older foster youth.[76] However, the analysis also found a number of barriers to effective youth services to be fairly common.

The most common identified barrier was inadequate or non-existent services, including service gaps related to life skills training, job skills training, substance abuse treatment, mental health treatment, general independent living services, inpatient mental health treatment services, services for pregnant or parenting teens, and independent living assessments. (Of the 45 final reports reviewed in this analysis 34 states reported the service gap barrier.) The second most common barrier (identified in 26 of the 45 reports analyzed) was a lack of placement resources for adolescents, including transitional living placements, homes for youth over 18, and homes for pregnant and parenting teens. Twenty-five of the 45 state final reports analyzed identified gaps in youth-specific training for staff and foster parents ranging from having no specialized adolescent training for any staff or foster parents to needing specialized training in assessing youths' needs. This training need was identified despite the fact that the majority of states received a positive rating relating to provision of training. Of the 45 states included in the analysis, 87% identified the need for additional training in adolescent issues despite the fact that they received an overall strength rating for their training program.

Inconsistency in services was found to be a barrier in a separate group of 25 states. This meant, for instance, that services related to life skills might be very strong in one part of the state but lacking in other areas of the state. In addition, in some places the service inconsistencies were found among contractors or providers within a specific region or area of the state. Finally, in 24 states the quality and consistency of independent living planning was identified as an issue. Specific problems included a lack of youth involvement in case planning, no independent living case plan completed, case plans not regularly updated, no transition planning, no youth assessment, poor placement matches, lack of individualized planning for youth in long-term care, and poor permanency goal selection.

A review of the initial 31 Program Improvement Plans approved for state implementation found that independent living and adolescent issues might not have been addressed. Researchers concluded that this might be the case because the language in the PIP was more general (services for all children not just adolescents) or because the specific concern was identified within an area for which the state might nonetheless have received a positive rating (e.g. quality of training program).[77]

GAO Report

A comprehensive report on the development and implementation of the CFCIP was produced by the GAO in November 2004.[78] The report was based on survey data collected from independent living coordinators in all 50 states, the District of Columbia, and Puerto Rico as well as review of the jurisdictions' CFCIP plans for FY2001 through FY2004. After the 1999 passage of P.L. 106-169, forty states reported expanded independent living services to youth younger than they had previously served and 36 states reported serving older youth. In addition, 45 states reported offering assistance with room and board to emancipated foster youth. According to officials in the states GAO visited for the study, funds were also used to improve the quality of existing independent living services, refocus the attention of their programs, or develop new services to assist youth of all ages in the programs.

The GAO report also raised concerns about the implementation of the CFCIP. The report indicates that states varied in the percentage of eligible youth served. In 2003, forty states reported serving between 10% and 100% of eligible youth, with one-third of the states serving less than half of eligible youth. GAO also found gaps in the availability of mental health services, mentoring services, and securing safe and suitable housing, particularly in rural areas. Further, although 49 states reported increased coordination with a number of federal, state, and local programs that can provide or supplement independent living services, child welfare administrators and youth interviewed by GAO said that they were unaware of the services. Finally, the lack of uniformity among the states' CFCIP five-year plans precluded using them at the state and federal level to monitor how well the programs serve eligible youth.

Implementation of the ETV Program

In 2007, the National Foster Care Coalition, in partnership with Casey Family Programs, a foundation supporting child welfare research and advocacy, reported on six states' (California, Maine, Montana, New York, North Carolina, and Wyoming) experiences with implementing the ETV program.[79] These states serve as few as 31 youth to as many as nearly 2,000 youth in a given school year, with average awards ranging from $2,950 to $4,318 for each youth. Three of the states (Montana, New York, and North Carolina) contract with a non-profit service provider to administer the ETV program, while the other three states administer the program through the independent living coordinator or state financial aid office.

The report describes the best practices employed by the states in implementing their ETV programs, as well as the challenges they have encountered.[80] One promising approach was the application process for some states, which involves a web application process that allows students and schools to view application and approval status and deadlines. Other states reported providing extensive promotion and outreach about the program through information sessions, annual teen conferences for foster youth, and mailings; and providing additional educational supports to youth through financial aid counseling, mentoring, tuition waivers, and scholarships. Contracting through a third-party was also identified as an important practice that has lent to more efficient administration of the voucher program. The report also identified youth feedback about the program as another important feature of the programs.

In addition, the report also identified barriers to successfully administering the ETV program. Some of the challenges include (1) meeting the demand for the program; (2) recruiting youth to the program and ensuring that these youth have sufficient support to remain in school; (3) managing awards for youth whom schools are unable to locate once awards are issued or awarding funds to youth who have dropped out; (4) tracking how voucher funds are spent; (5) conveying to university financial aid staff the rules associated with the ETV program; (6) maintaining the confidentiality of youth's foster care experiences; (7) connecting non-college-bound youth to vocational programs; and (8) meeting the needs of parenting youth, among others. The states indicated that they are working to address these challenges.

Chapter 5

OTHER FEDERAL SUPPORT FOR OLDER CURRENT AND FORMER FOSTER YOUTH

In addition to the federal programs under Title IV-E, other federal laws authorize some funding for service or assistance to older current and former foster youth. This section describes a Medicaid pathway for certain former foster youth; educational, workforce, and housing supports; and a grant to fund training for child welfare practitioners working with older foster youth and youth emancipating from care.

CHAFEE MEDICAID PATHWAY

In the Chafee Foster Care Independence Act, Congress encouraged states to provide Medicaid coverage to children who were aging out of the foster care system. The law created a new optional Chafee Medicaid eligibility pathway for "independent foster care adolescents;" this pathway is often called the "Chafee option."[81] The law further defined these adolescents as individuals under the age of 21 and who were in foster care under the responsibility of the state on their 18th birthday. Within this broadest category of independent foster care adolescents, the law permits states to restrict eligibility based on the youth's income or resources, and whether or not the youth had received Title IV-E funding.[82]

Based on a 2006 survey of state human service officials, 17 states (Arizona, California, Florida, Kansas, Indiana, Iowa, Massachusetts, Mississippi, Nevada, New Jersey, Oklahoma, Rhode Island, South Carolina, South Dakota, Texas, Utah, and Wyoming) have extended Medicaid coverage to youth eligible under P.L. 106-169; Missouri took up the Chafee Medicaid pathway in 2007.[83] According to the survey, four other states were considering adopting the option.

In all states, youth age 19 or younger with family incomes at or below 100% of the federal poverty limit (or up to 250% in some states) are eligible for Medicaid or State Children's Health Insurance Program (SCHIP). Youth ages 18 to 21 in foster care who do not qualify for Medicaid or SCHIP may be eligible for Medicaid coverage through the "Ribicoff" pathway, named for the late former senator, Abraham Ribicoff. Ribicoff youth must meet the income and resource requirements for the former Aid to Families with Dependent Children program but do not meet other categorical requirements for AFDC. Over half of all states have opted to provide coverage to former foster care youth through this pathway, although length of eligibility for coverage varies.[84] Finally, current and former foster youth may qualify for Medicaid through other eligibility pathways available to certain groups of adults, such as for pregnant women with family income equal to or less than 133% of the federal poverty limit (FPL) and some low-income adults with children.[85]

OTHER EDUCATIONAL SUPPORT[86]

As of July 2009, the College Cost Reduction Act (P.L. 110-84) will allow certain youth who have been in foster care to claim independent status when applying for federal financial aid. The act amended the definition of "independent student" in the Higher Education Act to include any child "who is an orphan, in foster care, or a ward of the court at any time when the individual is *13 years of age or older*" and "is an emancipated minor or is in legal guardianship as determined by a court of competent jurisdiction in the individual's state of legal residence." [87]The act does not specify the length of time that the child must have been in foster care or the reason for exiting care, to be eligible to claim independent status. The law first becomes effective for the 2009 to 2010 school year.

Students who claim independent status are typically able to access greater federal education assistance because they are exempt from including information about income and assets from their parents. An "independent" student's expected "family" contribution is the amount that the federal need analysis system determines should be contributed, based only on his or her available income (and assets, if applicable), as well as basic living expenses, federal income tax liability, and other expenses.

Workforce Support

Workforce Investment Act Programs

The Workforce Investment Act authorizes job training programs to unemployed and underemployed individuals through the Department of Labor (DOL). Two of these programs — Youth Activities and Job Corps — provide job training and related services to targeted low-income vulnerable populations, including foster youth.[88] The WIA Youth Activities program focuses on preventative strategies to help in-school youth stay in school and receive occupational skills, as well as on providing training and supportive services, such as assistance with child care, for out-of-school youth. Job Corps is an educational and vocational training program that helps students learn a trade, complete their GED, and secure employment.

To be eligible, foster youth must meet age and income criteria as defined under the act. Young people currently or formerly in foster care may participate in Youth Activities if they are ages 14 to 21, and in Job Corps if they are ages 16 to 24 (20% of participants must be ages 22 to 24).[89]

Foster Youth Workforce Demonstration Projects

The Workforce Investment Act authorizes funding for pilot programs.[90] Under this authority, the Department of Labor's Employment and Training Administration awarded grants to five states in FY2005 — California, Illinois, Michigan, New York, and Texas — to design and implement programs to improve the self sufficiency, education attainment, and employment skills of youth aging out of foster care. The purpose of the grant was to encourage states to develop best practices around serving foster youth in the workforce investment system, and integrate these practices across workforce investment boards across each state. The five states were required to target the programs to youth in areas with the largest foster care populations. These areas are Los Angeles, Chicago, Detroit, New York City, and Houston. DOL awarded each state $800,000 total for FY2005 and FY2006; states were required to provide 100% matching funds.

The programs have served over 1,000 youth, of whom 81% were ages 17 and older and nearly 60% were female at entry. About 71% of the youth were black.[91] The youth varied in their educational attainment at entry. Approximately 42% were in high school, 9% were in a post-secondary institution, 23% had dropped out of school, and 26% had graduated or obtained their GED but were not in a post-secondary institution. Further, at entry, about half of the youth had stable housing situations, just over one quarter were in independent living arrangements, and about 19% were in temporary housing or homeless. Some of the youth also faced additional barriers. About one of out five of the participants was currently or previously adjudicated or incarcerated and slightly less than 20% were parents.

The sites have differed in their recruitment and delivery strategies, but all have provided youth with academic instruction and support, preparation for and exposure to the work place, support in developing skills for self-sufficiency, and the supportive services youth need to succeed academically and in the work place. Approximately 46% of the youth have been enrolled in the programs at least seven to nine quarters; nearly 32% have been enrolled for four to six quarters; and about 22% have been enrolled for one to three quarters.

With their own funding, Casey Family Programs and its partners are conducting an evaluation of the program, and has reported preliminary results.[92] The evaluation found that in the second year of the grant (FY2006), grantees noted the challenges with serving the older foster care population because of the youth's mental health needs, low literacy levels, and anger management issues. All of the sites reported difficulty in forming partnerships that provide mentoring to youth. The sites also reported improvements in their programs. Partnerships with workforce investment boards and other public agencies were formed to provide youth with job training experience. At the state level, new relationships between the workforce agencies and the child welfare agencies have also been formed.

Further, nearly 45% of youth have obtained a GED or diploma, attended a post-secondary institution, or secured employment during their time in the program.[93] Youth who have been in the program longer were more likely to attain one of these outcomes than youth who were in the program for a shorter period of time. The evaluation has preliminarily concluded that no single agency can meet the needs of youth in foster care or aging out of care; case workers — who serve as social workers, counselors, mentors, and teachers — are highly valued by the youth; well-defined programs tended to be more successful in leveraging other services for youth; and the sites lacked consistency in they ways they defined and collected the data. However, the evaluation has also found that the sites made efforts to overcome the challenges with the data to better enable researchers to evaluate the youth's progress and outcomes.

The programs are continuing to serve eligible youth, with funding from Casey Family Programs and a 100% match by the states and other lead partners.

HOUSING SUPPORT

Family Unification Vouchers Program

Current and former foster youth may be eligible for housing services provided through programs administered by the Department of Housing and Urban Development's (HUD) Family Unification Vouchers program (FUP vouchers). The FUP vouchers were initially created in 1990 under P.L. 101-625 for families that qualify for Section 8 tenant-based assistance and for whom the lack of adequate housing is a primary factor in the separation, or threat of imminent separation, of children from their families or in preventing the reunification of the children with their families.[94] Amendments to the program in 2000 under P.L. 106-337 made youth ages 18 to 21 who left foster care at age 16 or older eligible for the vouchers. These youth are eligible for the vouchers for up to 18 months.

FUP vouchers were initially awarded from 1992 to 2001. Each award included five years of funding per voucher and the voucher's use was restricted to voucher-eligible families for those five years. At the end of those five years, public housing authorities (PHAs), which administer the vouchers, were eligible to convert FUP vouchers to regular Section 8 housing vouchers for low-income families. While the five-year use restrictions have expired for all family unification vouchers, some PHAs may have continued to use their original family unification vouchers for FUP-eligible families and some may have chosen to use some regular-purpose vouchers for FUP families. For FY2008, Congress provided $20 million for new FUP vouchers.[95] Congress has also specified that amounts made available under the FY2008 appropriations act and previous appropriations acts for Section 8 tenant-based rental assistance and used for non-elderly families or the FUP are to remain available for these purposes, to the extent practicable.

Other Support

Older current and former foster youth may be eligible for housing services and related supports through the Runaway and Homeless Youth program, administered by the U.S. Department of Health and Human Services.[96] In FY2007, over 1,000 of the 47,519 youth (2.1%) who used the program's Basic Center program or Transitional Living program, were living in foster homes at the time they entered the program.[97] That same year, of the 46,317 youth who exited the program, nearly 1,600 (3.4%) were placed in a foster home. (The number of youth who may have lived in other foster care settings before entering or at exit is not reported.) Youth transitioning out of foster care may also be eligible for select transitional living programs administered by HUD, though the programs do not specifically target these youth.[98]

Grant to Support Curriculum for Supervising Older Youth in Care

In FY2006, the U.S. Department of Health and Human Services awarded grants under Title IV-B of the Social Security Act (Promoting Safe and Stable Families) to fund the development of curriculum for child welfare supervisors and their staff who work with older youth in foster care. The grants were awarded through FY2008 to six entities: the Hunter College School of Social Work; Massachusetts Department of Social Services; San Francisco State University; University of Iowa; University of Houston; and University of Louisville Research Foundation. Each of the entities provide at least a 25% match to the federal grant award of (up to) $250,000 annually for three years.

The six entities are to develop, implement, evaluate, and disseminate a training curriculum for public child welfare agency supervisors.[99] The purpose of the curriculum is to strengthen supervision of staff interactions with older youth in care and/or in independent living programs, and to ensure that staff adequately: (1) assess a youth's readiness for independent living services, support, and training; (2) identify culturally competent independent living program services and activities; (3) utilize positive youth development principles for involving youth in decisionmaking, implementation, and evaluation of training and program activities;[100] (4) identify areas of stress and its impact on youth in foster care; (5) work with youth to help them deal with crisis situations and to assess the results of the intervention; (6) work with youth to develop and maintain permanent connections; and (7) collaborate with both inter-and intra-agency resource people to achieve positive outcomes for youth transitioning to adulthood. Each entity must conduct an evaluation of the project, either in-house or by contracting with a third-party evaluator.

One of the grantees, the National Resource Center for Family-Centered Practice and Permanency Planning (NRCFCPPP) at the Hunter College School of Social Work, in partnership with the National Foster Care Coalition and Child Welfare League of America, has developed and provided training on curriculum to three state/city partners (Mississippi, New York City, and Oregon) based on a learning circle model. Supervisors in the three jurisdictions participate in six learning circle sessions with topics on youth development and older youth in care, such as creating permanent connections for youth and caring adults, relating to youth as resources rather than recipients of child welfare services, and involving a diverse array of stakeholders in the development of a comprehensive set of services and supports for youth transitioning out of care. At each learning circle session, staff are required to develop action plans to guide their work. In turn, supervisors use the materials and discussion guide from the learning circle sessions (and available on the university's website) to train their staff one-on-one, in group settings, or at unit meetings or staff retreats. The evaluation for the program is being conducted by the Hunter College School of Social Work. During the first two years of the program, the school has conducted a process evaluation to make needed adjustments to the curriculum content and training delivery methods. The school is in the process of conducting an outcome evaluation to measure the effectiveness of the curriculum.

The remainder of this book discusses issues related to the federal role in providing support to current and former older foster youth, as well as related pending legislation and hearings in the 110th Congress.

Chapter 6

ISSUES[101]

FOSTER CARE FOR YOUTH AGES 18 AND OLDER

As discussed above, some states report allowing youth to remain in care after their 18th birthdays under certain conditions, such as the youth attending a college or university (see Appendix B and Appendix C). Yet few states appear to actually facilitate youth staying in care. In light of the negative outcomes that young people often experience upon emancipating, policymakers and child welfare practitioners have raised concerns about policies requiring youth to leave foster care custody on their 18th birthday. On July 12, 2007, the Subcommittee on Income Security and Family Support of the House Ways and Means Committee held a hearing to highlight the challenges youth aging out of care face as they transition to adulthood.[102] The witnesses — child welfare practitioners and researchers and youth — explained that foster care services end abruptly for many older youth in care when they reach age 18. One practitioner said that a county child welfare agency has begun buying luggage for youth aging out to replace the garbage bags they were using to haul their few belongings. He went on to say that this practice is merely a band-aid and does not address the bigger issue that youth often lack housing and other support upon aging out of care.

While there is a paucity of information about outcomes for youth who remain in care after age 18, findings from wave 3 of the Midwest Evaluation (see above for more information) suggests that youth who remain in care as late as age 20 tend to experience an easier transition to adulthood than their counterparts who emancipate at age 18. The study examined outcomes for former foster youth in three states —Illinois, Iowa, and Wisconsin. These three states offer a natural experiment for comparing youth outcomes: Iowa and Wisconsin emancipated nearly all of the foster youth in the study by age 21, while approximately three-fourths of foster youth in Illinois who reached age 18 in care remained under the custody of the state until age 21.[103] The study found that the young adults from Illinois were almost two times as likely to have ever attended college than their peers in Iowa and Wisconsin; and after controlling for observed differences (i.e., gender, race/ethnicity, age at most recent entry into care, number of prior placements, among other characteristics), Illinois youth were four times as likely to have ever attended college and approximately 3.5 times as likely to complete one year of college.[104]

Further, remaining in care appears to be associated with higher earnings and delayed pregnancy. The study found that while the young people in Illinois were less likely to be employed, due likely to being in school, each additional year in care after age 18 was associated with a $470 increase in annual earnings. Annual earnings for youth who remained in care longer increased by $924 after controlling for certain characteristics of the young adults (measured at baseline) that are likely to affect later earnings (e.g., work history, education attainment, mental health problems, and criminal behavior), as well as unobserved characteristics. Further, young people in Illinois were 38% less likely to become pregnant between ages 17 and 19. Although there was a reduction in the risk of pregnancy after age 19 for youth in care compared to their counterparts, this difference was not statistically significant.

Eligible Placement Setting

Extending federal foster care for youth age 18 and older raises the question about whether states should be reimbursed for youth who do not live in traditional foster care settings. A state may only claim some federal reimbursement for foster care maintenance payments made on behalf of a child if that child is placed in an "eligible setting" (Section 472(a)(2)(c)). Under current law, this is defined to include a licensed foster family home or "child-care institution," including group homes or other congregate care facilities (Section 472(c)).[105]

Further, a foster care maintenance payment may only be made to a foster family home or to a child care institution. No foster care maintenance payment may be made directly to an older youth who is in a supervised independent living situation or to help an older foster youth pay his or her rent.[106] Yet, many older foster youth — even those who remain in state care — do not live with a foster family (or in a child care institution) but instead live in supervised independent living quarters or with relatives. The Midwest Evaluation of the Adult Functioning of Former Foster Youth study found that at age 19, half (50%) of the surveyed youth (some of whom were in foster care) lived in supervised independent living settings; about 20% lived in foster family homes; and 19% lived with relatives. The remainder lived in group quarters (7%) or other settings (4%).[107]

Judicial Oversight

Foster care for youth ages 18 and over also raises questions about the role of the juvenile court in supervising the cases, given that these youth are legally adults. Appendix C describes foster care programs in four states — Illinois, New York, Vermont, and Washington, D.C. — for youth age 18 and older. While youth in Vermont do not remain under court jurisdiction, youth ages 18 and older in care in Illinois, New York, and Washington, D.C. continue to be wards of the state. The juvenile courts retain jurisdiction and social workers make routine visits to assist youth in achieving their case goal (usually independent living).

Relevant Legislation

Four bills pending in the 110th Congress propose to require or encourage states to provide foster care to youth ages 18 and older.

The Foster Care Continuing Opportunities Act (S. 1512, introduced by Senator Barbara Boxer), and the Invest in KIDS Act (H.R. 5466, introduced by Representative Jim McDermott) propose to give states *the option* of seeking federal reimbursement under the Title IV-E foster care program for otherwise eligible youth who elect to remain in foster care until their 21st birthday or at state option until their 19th or 20th birthday.[108] By contrast, the Place to Call Home Act (H.R. 3409, introduced by Representative Ruben Hinojosa) and Reconnecting Youth to Prevent Homelessness Act of 2007 (H.R. 4208, introduced by Representative Shelley Berkeley and S. 2560, introduced by Senator John Kerry), would *require* states to make foster care maintenance payments on behalf of eligible youth who choose to be in foster care after their 18th birthday (but not beyond their 21st birthday).

Under H.R. 3409 and H.R. 4208/S. 2560, as part of choosing to remain in foster care, a youth would be required to enter into a written "voluntary placement agreement," that would be binding on both the youth and the state agency, and would specify the youth's legal status as well as the rights and obligations of both the state agency and the youth while he or she remains in foster care. Because they provide access to additional federal funds for foster care, S. 1512 and H.R. 5466 (as well as H.R. 3409 and H.R. 4208/S. 2560) might encourage more states to both permit youth to elect to remain in foster care and to facilitate this longer-term placement. Ultimately, the decision to remain in care, however, would appear to remain with the youth who would otherwise be exiting care.

H.R. 3409 and H.R. 4208/S. 2560 would also amend current law to explicitly provide that foster care maintenance payments would be eligible for federal reimbursement even if they are made to cover the cost of a licensed supportive housing facility or to cover the rent for a youth's apartment. H.R. 5466 would amend current law to permit youth placed in a "supervised setting" but who are living independently to remain eligible for the payments, in accordance with any conditions established by the HHS Secretary by regulation with regard to such a setting. S. 1512 does not address the issue of "eligible placement setting."

The four bills are silent about the role of the juvenile court in retaining oversight of young people in care who are legally adults.

PERMANENCY

Several federal provisions require the state child welfare agency to assist youth in planning for their transition from foster care to independent living. Currently under Title IV-E, a case review must be conducted not less often than every six months by a judge or an administrative review panel and at least once every 12 months by a judge who must consider the child's permanency plan (Section 475(5)). Specific case plan and case review procedures pertain to older youth in care. As described above, for a child age 16 or older, the written case plan must include a description of the programs and services that will help the child prepare for the transition to independent living (Section 475(1)(D)), and the permanency plan hearing must consider "the services needed to assist the child to make the transition from foster care to independent living" (Section 475(5)(C)).

Despite these federal protections to ensure that child welfare agencies help youth plan for their future, child welfare practitioners and young people in care continue to advocate for additional policies that improve the transition to adulthood by encouraging strong, permanent connections to caring adults.[109] A youth at the July 12, 2007 hearing on youth aging out of foster care, said that permanency can help young people avoid poor outcomes. He said, "[p]ermanency is having someone there to help you when you need it, someone you don't need an appointment to talk to. Permanency is having someone to lean on for support when obstacles come your way."[110]

In some jurisdictions, the child welfare agency plays an active role to ensure permanent connections for youth aging out. The National Governors Association reports that connections may include (1) connecting youth to family members such as grandparents and aunts; (2) establishing legal guardianship or some other permanent arrangement with a caring adult in the youth's life; (3) helping youth develop relationships with caring adults such as teachers or mentors; (4) when appropriate, exploring the viability of reunification with biological parents; and (5) assisting youth, up to their early twenties, pursue adoptive relationships with adults.[111]

Some states, like California, require child welfare agencies to help youth build permanent connections. County child welfare agencies in California are to provide "assistance in maintaining relationships with individuals who are important to a child who has been in out-of-home placement in a group home for six months or longer from the date the child entered foster care, based on the child's best interests" (California Welfare and Institutions Code Sec. 391). In Alameda County, east of San Francisco, the county child welfare agency conducts a transition conference for each youth aging out of care.[112] At the meeting, stakeholders in the youth's life (e.g., social worker, mental health counselor, family member(s), family friend(s), teacher(s), and foster parent(s)) help the youth establish a plan for transitioning to adulthood. The conference is intended to assist youth in planning their future and to ensure that one or more of these stakeholders will provide the youth with emotional and other support for the years following his or her emancipation from care.

Relevant Legislation

H.R. 3409 and H.R. 4208/S. 2560 would require that a permanency hearing for a youth transitioning from foster care to a planned permanent living arrangement or independent living is held in a family or juvenile court or another court (including a tribal court) of competent jurisdiction. Further, the bills would require states to include in the written case plans for children aging out of foster care, the steps taken to ensure that a child has a permanent living arrangement if they emancipate from care; in the case of a child age 17 or over with a permanency goal of emancipation, the state must provide documentation of the permanent living arrangement. H.R. 3409 and H.R. 4208/S. 2560 would provide that where appropriate, the state is to include in the written case plan for a child who is 14 years old in care, a description of the programs and services that will facilitate his or her transition from foster care to independent living. The plan must discuss the appropriateness of the services that have been provided to the child as well as documentation of the steps the agency is taking to find a permanent placement with a family or other adult connection for the youth, as well as a permanent living arrangement. In addition, the bills would require that the permanency hearing review all documentation of the efforts to secure a permanent living arrangement for the child upon emancipating.

H.R. 3409 and H.R. 4208/S. 2560 would further require that the state inform all children leaving care of the full range of available financial, housing, counseling, health, and other services for which the youth is eligible. In addition, the HHS Secretary would be required to provide for the "efficient distribution" to states and local areas information about federal programs, other than the CFCIP, that can assist youth in the transition to self-sufficiency and how to access the services under these programs.

Finally, the Foster Care Mentoring Act (S. 379), also pending in the 110th Congress, would permanently authorize funding, under Title IV-B of the Social Security Act, for grants to provide mentoring to children in foster care. The grants would be awarded by HHS to states (or to a political subdivision of the state if it serves a "substantial number" of youth in foster care) to support, establish, and expand networks of public and private community entities to provide this mentoring. Successful applicants would be eligible to receive a maximum of $600,000 annually, would be required to spend no less than 50% of the federal grant funds for training (and no more than 10% on program administration), and would need to provide matching funds (in cash or in kind) of 25%. The bill would authorize $15 million for this grant program in each of FY2008 and FY2009 and "such sums as may be necessary" in every following fiscal year.

Housing

Among the most often cited concerns for youth aging out of foster care is the lack of adequate and affordable housing.[113] In the Northwest Foster Care Alumni Study of former foster youth ages 20 to 33, about one out of five (22.5%) reported being homeless for one day or more within a year of leaving foster care.[114] (This figure is slightly higher than the 18% of 21-year-olds in the Midwest Evaluation who reported being homeless at least once since existing care.[115]) The housing status of former foster youth is often affected by relationships, education, and employment.[116] Reciprocally, youth who lack housing may have difficulty staying in school and/or maintaining employment.

While the CFCIP authorizes states to spend up to 30% of their allotment on room and board for youth ages 18 to 21, child welfare researchers point out that these funds alone cannot adequately cover the cost of housing for many youth.[117] However, several jurisdictions appear to be using innovative strategies to connect youth to both temporary and permanent housing. These strategies have involved providing a continuum of housing that allows the youth to choose from various housing options (e.g., scattered-site apartments, supervised apartments, shared homes with multiple youth, host homes, and boarding homes); permitting youth to have multiple opportunities to remain in a housing program even if they fail to meet the terms of a housing agreement or lease; and providing housing to former foster youth with mental health issues and other barriers to living independently.[118]

Another related issue is temporary housing for youth in college who are unable to stay with family or friends over school breaks. One of the youth witnesses at the July 12, 2007 hearing described the difficulty in college with finding housing when her dorm was closed: "I waited in limbo for a friend to extend an invitation as I wondered where I would go for holidays and school breaks."[119] Some states require public universities to provide housing for these youth. For example, California law requires that the California State University system and the community college system, "review housing issues for those emancipated foster youth living in college dormitories to ensure basic housing during the regular academic school year, including vacations and holidays other than summer break" (California Education Code Sec. 89342).

Relevant Legislation

H.R. 3409 would amend the Family Unification Vouchers program, described above, to make former foster youth eligible for the vouchers until age 25. The bill would also eliminate the requirement that these youth may be eligible for the vouchers no more than 18 months.

The College Opportunity and Affordability Act of 2007 (H.R. 4137, introduced by Representative George Miller) and the Higher Education Amendments of 2007 (S. 1642, introduced by Senator Edward Kennedy) would authorize funding from federal higher education programs (Fund for the Improvement of Postsecondary Education in S. 1642 and Student Support Services in H.R. 4137) to be used to establish demonstration projects that provide comprehensive support services for homeless students or students who were in foster care until the age of 18, to enroll and succeed in postsecondary education, including providing housing to students when housing at an educational institution is closed or unavailable to other students.

RUNAWAY YOUTH[120]

A child is considered missing from foster care if she or he is not in the physical custody of the child welfare agency or the institution or person with whom the child has been placed, due to (1) the child leaving voluntarily without permission (i.e., runaways); (2) the family or nonfamily member removing the child, either voluntarily or involuntarily, without permission (i.e., abductions); or (3) a lack of oversight by the child welfare agency.[121] The majority of children known to be missing from foster care are runaways. According to the U.S. Department of Health and Human Services, on the last day of FY2006, approximately 12,000 (2%) of the 510,000 children in foster care had run away, and another 5,049 had exited the system as runaways (because they were old enough to emancipate and were on runaway status at the age of emancipation).[122] As shown in the Foster Care Dynamics study, discussed above, most runaways tend to be teenagers.

A study of youth in the Midwest who ran away from foster care between 1993 and 2003 found that the average likelihood of an individual running away from foster care placements increased over this time period.[123] Youth questioned about their runaway experiences cited three primary reasons why they ran from foster care: (1) to reconnect or stay connected to their biological families even if they recognized that their families were neither healthy nor safe; (2) to express their autonomy and find normalcy among sometimes chaotic events; and (3) to maintain surrogate family relationships with non-family members. Youth in the study were more likely than their foster care peers to abuse drugs and to have certain mental health disorders.

No federal laws specifically address the issue of children missing from foster care, including runaways. However, Titles IV-B and IV-E of the Social Security Act require state child welfare agencies to monitor and provide for the safety and well-being of children in out-of-home care. Under Section 471 (Title IV-E), states are eligible for federal foster care and adoption assistance support if, among other requirements, they develop a written case plan for each child regardless of IV-E eligibility (Section 475(1) of Title IV-E). The case plan must discuss the safety and appropriateness of the placement and a plan for assuring that the child receives safe and proper care.

States must also develop a system to review, no less than every six months, the status of the child's case plan.[124] Also, under Section 471, states must conduct background checks for prospective foster parents before approving a placement. Finally, under Section 422 (Title IV-B), states must ensure that children in foster care are visited by their caseworkers on a monthly basis and that the majority of the visits occur in the child's residence.

Relevant Legislation

H.R. 3409 and H.R. 4208/S. 2560, discussed above, propose amending Section 471 of the Social Security Act to require states to include in their foster care and adoption assistance plans a description of their written policies and procedures designed to reduce the incidence of children missing or running away from foster care and to locate and return these children to foster care placements.

Use of Chafee Education and Training Vouchers

Under current law, the Chafee Education and Training Voucher program is available for youth who are eligible for CFCIP services or were adopted from foster care after 16 years of age. Youth may continue to be eligible for the program until age 23, but only if they received a voucher at age 21. According to some child welfare practitioners and researchers, the program may provide a disincentive to establish permanent relationships for youth through adoption because foster youth adopted before the age of 16 would not be eligible for the vouchers.[125] Further, studies of older foster youth and youth leaving care show that many do not attend college directly after emancipating.[126] Therefore, former foster youth who start college after age 21 would be ineligible for the program.

Relevant Legislation

The pending Kinship Caregiver Support Act (H.R. 2188, introduced by Representative Danny Davis), H.R. 3409, and H.R. 4208/S. 5260 would permit youth who left foster care at age 14 or older under a kinship guardianship arrangement or because they were adopted to be eligible for ETVs.

Another bill could shift the target population and purpose of the education and training voucher program. The School Choice for Foster Kids Act (H.R. 4311, introduced by Representative Michele Bachmann) would enable the education and training voucher program to fund the education of foster children — of any age —at private schools and to transport children to their original school, even if the school is outside of their immediate area of placement.[127] At a June 19, 2007 hearing on disconnected and disadvantaged youth, conducted by the House Ways and Means Subcommittee on Income Security and Family Support, Representative Michele Bachmann and Dan Lips with the Heritage Foundation testified that foster children face several educational challenges, and that additional educational support is needed for the population.[128] They cited that foster children are more likely to experience multiple school placements, which may lead to gaps in learning and lowered educational attainment, and that a disproportionate share of foster youth have Individualized Education Plans (IEPs) to accommodate their learning disabilities. They suggested that funds from the ETV program could be used to improve the educational outcomes of children and youth in care.

Medicaid Coverage for Youth Aging out of Care

The Midwest Evaluation indicated that about half of all youth who were in foster care up to age 17 or 18 (primarily in Iowa and Wisconsin) did not have health insurance when they were surveyed one year after leaving care, at age 18 or 19.[129] These studies also suggest that these youth need health and mental health services. Current and former foster youth at age 18 or 19 tended to describe their overall health less favorably than youth in the general population and were more likely to report that health conditions limited their ability to engage in moderate activity.[130] They also reported more visits to the emergency room and more hospitalizations during the past five years. These youth were also more likely to report being hospitalized for drug use or emotional problems.[131]

Some child welfare advocates support expansion of the Chafee Medicaid option to provide Medicaid for youth who age out of foster care. According to the American Public Human Services Association (APHSA), which surveyed five states (California, Florida, Iowa, South Carolina, and Texas) on their Medicaid coverage to youth who have aged out, the costs of the program vary depending on the type of program offered, but even the highest cost — $350 per youth per month (South Carolina) — is affordable.[132]

Relevant Legislation

Five bills pending in the 110th Congress would expand Medicaid coverage to former foster youth, either directly or indirectly. The Medicaid Foster Care Coverage Act of 2007 (H.R. 1376, introduced by Representative Dennis Cardoza), proposes to provide mandatory Medicaid coverage to any youth under the age of 21 who was in foster care on his or her 18th birthday and without regard to whether the youth chose to remain in foster care. H.R. 3409 would make this same change, however, it would provide that this mandatory Medicaid coverage category would continue until the youth reached the age of 25. Both H.R. 1376 and H.R. 3409 would *permit* states to limit this mandatory coverage to those youth who were in foster care on their 18th birthday who meet certain income and asset criteria, were previously Title IV-E eligible, or who received CFCIP services. H.R. 3409 specifies that those CFCIP services may have been received up until the youth's 25th birthday.

S. 1512, H.R. 5466, and H.R. 4207 would not make any statutory changes to Medicaid eligibility rules. However, because children eligible for Title IV-E foster care maintenance payments are automatically eligible for Medicaid coverage,[133] by expanding the population of children eligible for those payments, the bill also would extend mandatory[134] Medicaid coverage to any youth who elects to remain in foster care after his or her 18th birthday (provided that youth meets all the other Title IV-E eligibility criteria) for as long as the youth remained in foster care or until his or her 21st birthday (whichever came first).

The Risk of Becoming Disconnected

In its February 2008 report on disconnected youth, the U.S. Government Accountability Office defined this population as youth ages 14 to 24 who are not in school and not working, or lack family or other support networks.[135] According to the report, some of these young people may have become disconnected from education and employment through incarceration, aging out of foster care, dropping out of high school, or homelessness. At the June 19, 2007 hearing on disconnected and disadvantaged youth, witnesses also identified emancipation from care as a potential pathway to becoming disconnected.[136] Chairman Jim McDermott and Ranking Member Jerry Weller expressed that policymakers have become increasingly concerned about the disconnected youth population because of the potential negative outcomes they may experience, and that they may lack the ability to compete for jobs and earn a living wage in the increasingly global economy.

An analysis by the Congressional Research Service of the U.S. Census Bureau's Current Population Survey (CPS) data uses a definition of disconnectedness to include noninstitutionalized youth ages 16 through 24 who did not work anytime during a previous year due primarily to a reason other than school and were presently (usually March or April of the current year) not working or in school. Thus, otherwise young people who are married without children (to a connected or disconnected partner) or are cohabiting with or without children meet the definition of being disconnected. Approximately 1.8 million youth — or 4.9% of all youth — ages 16 to 24 met this criteria.[137] Although not directly comparable, the Midwest Evaluation found that at age 19, 23.2% of females and 45.3% of males in foster care or who had aged out, met one of the researchers' definition of disconnected, which excluded parenting youth who were not working or in school. At age 21, 11.8% of females and 36.6% of males were disconnected.

Relevant Legislation

Several pending bills would provide educational and other supports to older youth in care and youth aging out of care to strengthen their connections to education and employment.

Education Support through the TRIO and GEARUP Programs

The federal TRIO and GEARUP programs seek to increase disadvantaged students' high school completion and enrollment in higher education by providing a variety of academic, counseling, and college preparatory services.[138] H.R. 4137[139] and S. 1642 would allow or require TRIO funds (specifically for the Educational Opportunity Centers, Student Support, Talent Search, and Upward Bound programs) and GEARUP funds (for H.R. 4137 only) to be used for programs and activities that are designed for students who are in foster care or are aging out of foster care and to recruit these youth to Educational Opportunity Centers, which provide counseling and information on college admissions to qualified adults who want to enter or continue a program of postsecondary education. Further, H.R. 4137 would require applicants for TRIO funds to coordinate with other programs for disadvantaged students (regardless of the funding source of such programs) that provide services to foster youth. H.R. 4137 would also require grantees seeking funding under the GEARUP program for early intervention activities (e.g., comprehensive mentoring, counseling, outreach, and support services to participating students) to treat as a priority any student in preschool through grade 12 (and may consider a student in the first year of attendance at an institution), who is in foster care or a homeless or unaccompanied youth, as defined by the McKinney-Vento Homeless Assistance Act.

Financial Support

The Focusing Investments and Resources for a Safe Transition Act (S. 2341, introduced by Senator Hillary Rodham Clinton), would amend the Child Abuse Prevention and Treatment Act (CAPTA) to authorize financial support and financial counseling for youth aging out of foster care. The bill would authorize "such sums as may be necessary" for FY2008 through FY2012 to permit HHS to make competitive grants to states (or state partners) to establish individual development accounts (IDAs) for foster youth, including those in kinship or guardianship placements, and youth transitioning from foster care. IDAs are savings accounts to help low-income families and persons save for specified purposes, usually education, purchase of a home, or to start a business.[140] To be eligible for the funds, states would be required to submit a plan to HHS that describes how an IDA program would best suit the current and future needs of the state's foster youth, enable foster youth to achieve self support after leaving foster care, and establish public or private partnerships to create a pool of funding from which foster care deposits can be matched (not by more than $2 for every $1 deposited by a youth).

Funds saved in this account could be used by a youth for housing, education, vocational training, to operate a business or to purchase a car (though youth must expend funds on the first three purposes before being permitted to spend funds on these last two) and, at the option of the state, for purchase of work-related items or car insurance to assist the individual in becoming independent. Youth would be eligible to withdraw the funds upon reaching age 18 and completing money management training. Any savings accumulated in an account during the period which a youth maintains or makes contributions to the account would not be counted for determining eligibility for other benefits under federal law (other than the Internal Revenue Code of 1986). S. 2341 would also require the HHS Secretary to conduct evaluations of the program and prepare reports to Congress that provide information about how youth spent the funds, how the state program impacted quality of life indicators after the youth withdrew the funds, the effectiveness of the money management training, and recommendations on strengthening the program.

The bill appears to be based in part on the Jim Casey Youth Opportunities Initiative's Opportunity Passport, a program to increase the financial assets and literacy of youth transitioning from foster care. The program, implemented in ten cities throughout the country, assists current and former foster youth build financial assets, including through a matched savings account. Of the 1,740 youth who participated in the program as of December 31, 2006, one quarter have used their savings to draw match funds for an approved asset; in most cases, the assets were a car (58%), housing (28%), and educational expenses (28%).[141] Youth who purchased assets tend to be older and no longer in care compared to their counterparts who did not purchase assets.

APPENDIX A. OUTCOMES FOR YOUNG ADULTS FORMERLY IN FOSTER CARE

Table A-1. Comparison of Outcome Domains between Young Adults in the Midwest Study and Young Adults in the Add Health Study

Outcome	Midwest Evaluation (Wave 3) - Former Foster Youth at Age 21 or 22	Add Health - Youth Surveyed at Age 21
Current Living Arrangement (totals to 100% across rows in each column)		
Lives in own place	44.3%	46.9%
Lives with biological parent(s)	7.6%	41.0%
Lives with other relative	16.8%	3.0%
Lives with non-relative foster parent(s)	5.6%	0%
Lives with spouse/partner	6.6%	0.4%
Lives with a friend	6.5%	1.2%
Lives in group quarters (e.g., dormitories, barracks)	3.1%	6.7%
Imprisoned or in jail	7.1%	
Other living arrangement	2.4%	0.8%
Highest Educational Attainment (totals to 100 across rows in each column)		
No high school diploma or GED	23.0%	10.8%
High school diploma only	37.6%	29.7%
GED only	9.7%	6.6%
One or more years of college, but no degree	27.9%	43.0%
Two-year college degree	1.9%	8.1%
Four-year college degree	—	1.7%
Graduate school	—	0.1%
Employment, Income, and Assets		
Ever held a job	95.1%	96.9%

Table A-1. (Continued).

Outcome	Midwest Evaluation (Wave 3) - Former Foster Youth at Age 21 or 22	Add Health - Youth Surveyed at Age 21
Currently employed[a] (nonincarcerated youth only)	44.5%	63.9%
Mean hourly wage[a]	$8.85	$9.99
Mean income[a]	$8,914	$12,728
Any savings/checking account[a]	51.9%	80.7%
Owns a vehicle[a]	39.1%	73.0%
Economic Hardships		
Not enough to pay rent[a]	26.5%	8.6%
Not enough money to pay utility bill[a]	26.5%	10.9%
Gas or electricity shut off	8.3%	6.1%
Evicted[a]	8.3%	1.4%
Receipt of food stamps[a]	50.2% - females	6.3% - females
	9.9% - males	0% - males
Receipt of TANF	8.8% - females[a]	7.5% - females[a]
	0.3% - males	0% - males
Health and Access to Health Care Services		
Description of general health as fair[a]	12.2%	4.0%
Description of general health as poor[a]	2.0%	0.3%
Health conditions or disability limits daily activities[a]	11.0%	4.7%
Has medical insurance[a]	50.7%	76.0%
Did not receive needed medical care[a]	17.9%	24.1%
Received psychological or emotional counseling[a]	10.5%	7.3%
Attended substance abuse treatment program	3.6%	2.3%
Sexual Behaviors and Pregnancy		
Age at first intercourse	16.0 - female	16.0 - female
	15.0 - male	16.0 - male
Had sexual intercourse in the past year	78.2 % - females	83.0% - females
	71.2% - males[a]	81.2% - males[a]
Used birth control all or most	60.4% - females	69.4% - females
of the time in the past year[a]	56.8% - males	67.9% - males
Ever paid by someone to have sex[a]	7.3% - females	1.8% - females
	14.0% - males	6.0% - males
Ever pregnant (females only)[a]	70.9%	33.8%

Outcome	Midwest Evaluation (Wave 3) - Former Foster Youth at Age 21 or 22	Add Health - Youth Surveyed at Age 21
Impregnated partner (males only)[a]	49.2%	19.2%
Relationships and Parenting		
Ever married[a]	11.5 % - females	17.9% - females
	5.1% - males	10.1% - males
Currently married	11.1% - females	16.2% - females
	4.3% - males[a]	8.6% - males[a]
Currently cohabiting	22.6% - females	16.7% - females
	18.0% - males	13.5% - males
At least one living child[a]	56.1% - females	23.5% - females
	30.2% - males	11.5%- males
Criminal Justice[b]		
Ever arrested[a]	56.7% - females	4.3% - females
	79.4% - males	20.1% - males
Ever convicted[a]	24.5% - females	1.3% - females
	52.6% - males	12.1% - males
Transition to Adulthood and Orientation Toward the Future		
Became socially mature[a]	66.2% - faster than others 28.3% - about the same rate as others 5.5% - slower than others	63.7% - faster than others 8.0% - about the same rate as others 28.3% - slower than others
Took on adult responsibilities[a]	67.9% - faster than others 26.5% - about the same rate as others 5.6% - slower than others	68.2% - faster than others 7.3% - about the same rate as others 24.5% - slower than others
Thinks of self as an adult[a]	Never or seldom - 4.9% Sometimes - 8.7% Most or all of the time - 86.3%	Never or seldom - 9.3% Sometimes - 18.2% Most or all of the time - 72.6%
Will live to 35 (mean score based on 1 to 5 scale, with 1 being almost no chance to 5 being almost certain)[a]	4.4	4.7
Will be married within next 10 years (mean score based on 1 to 5 scale, with 1 being almost no chance to 5 being almost certain)[a]	3.4	3.9
Will have a middle-class income by age 30 (mean score based on 1 to 5 scale, with 1 being almost no chance to 5 being almost certain)[a]	3.6	4.1

Table A-1. (Continued).

Outcome	Midwest Evaluation (Wave 3) - Former Foster Youth at Age 21 or 22	Add Health - Youth Surveyed at Age 21
Mentoring		
Maintained a positive relationship with a caring adult since age 14[a]	60.3%	77.4%
Closeness to mentor[a]	13.8% - not at all to a little close 13.6% - somewhat close 72.9% - very or quite close	22.4% - not at all to a little close 24.0% - somewhat close 53.6% - very or quite close

Source: Congressional Research Service presentation of data in Mark E. Courtney et al., *Midwest Evaluation of the Adult Functioning of Former Foster Youth: Outcomes at Age 21*, Chapin Hall Center for Children, University of Chicago, Dec. 2007.

Note: The Midwest Evaluation has tracked the outcomes of foster youth at age 17 and when they have aged out of care at ages 19 and 21 (some of these youth remained in care until age 21). For each of the three data collection waves, wherever possible, researchers asked the same questions that were taken directly from the National Longitudinal Survey of Adolescent Health ("Add Health"), a nationally representative survey that tracks a cohort of youth over time

[a]. Indicates that the difference between the youth in the Midwest Evaluation and youth in the Adolescent Heath Survey is statistically signficant.

[b]. The Add Health figures reflect arrests and convictions since age 18. The Midwest Study figures represent arrests and convictions since the wave 1 interview, when 62% of the young adults in the wave 3 sample were still 17 years old. Data for "ever arrested" were missing for 10 young men and 22 young women and data for "ever convicted" were missing for 24 young men and 35 young women.

APPENDIX B. MAXIMUM AGE AND CONDITIONS FOR YOUTH REMAINING IN FOSTER CARE AFTER THEIR 18TH BIRTHDAY

Table B-1. National Child Welfare Resource Center for Youth Development Survey of States, 2006

State	Age at Which Youth May No Longer Remain in Foster Care	Conditions for Extended Foster Care (after 18th Birthday)
Alabama	21	If youth is in school and in approved placement.
Alaska	20	Youth may remain in care until age 19, without the youth's consent and in care until age 20 with the youth's consent; under both conditions, placement must be in the "best interest" of the child.
Arizona	21	Youth who leave care at age 18 or older may return to care at any time before their 21st birthday.
Arkansas	21	If the youth is pursuing a post-secondary education.
California	19	If the youth will graduate from high school before 19th birthday.
Colorado	23	*Did not respond to inquiry about conditions for remaining in care.*
Connecticut	23	*Did not respond to inquiry about conditions for remaining in care.*
Delaware	21	If the youth is attending school and wishes to remain in placement.

Table B-1. (Continued).

State	Age at Which Youth May No Longer Remain in Foster Care	Conditions for Extended Foster Care (after 18th Birthday)
District of Columbia[a]	21	Court must approve request to terminate care before 21.
Florida	18	The youth may request an extension of jurisdiction until age19 so that the courts may monitor the provision of independent living services. They are not considered "in foster care" but are under the supervision of the court.
Georgia	21	If the youth has an educational plan in place and signs an agreement with the resource provider and the county. Youth may remain until age 21½ if the additional six months in care will enable youth to complete an educational program.
Hawaii		*Did not respond to survey.*
Idaho	21	If the youth is still in school and working on an independent living plan that includes continuing education, employment, and self-sufficiency skills.
Illinois[a]	21	*Did not respond to inquiry about conditions for remaining in care.*
Indiana	21	If the youth is in high school and making an effortto graduate;
		or if youth has been accepted into housing services (for
		developmental disabilities) and awaiting placement.
Iowa	20	If the youth is in high school or working toward a GED.
Kansas	21	*Did not respond to inquiry about conditions for remaining in care.*
Kentucky	21	Youth may request to extend commitment or reinstate commitment to remain in care. It must be approved by the court.
Louisiana		*Did not respond to survey.*
Maine	21	If the youth needs care and support for educational, social, or physical reasons.
Maryland	21	If the youth has documented special needs, is employed, and/or enrolled in a formal education program.

State	Age at Which Youth May No Longer Remain in Foster Care	Conditions for Extended Foster Care (after 18th Birthday)
Massachusetts	22	If the youth pursues education or vocational training and complies with their service plan.
Michigan	20	Foster care maintenance payments are available for former foster youth who have reached age 19, yet are still in a school or training program, regardless of whether youth are in family foster care or independent living settings.
Minnesota	21	Youth can request foster care benefits up to age 21, then must enter into a plan with a county social worker regarding education, employment, etc. If the county denies the request, the child, parent(s) or foster parent may appeal to the state Department of Human Services.
Mississippi		*Did not respond to survey.*
Missouri	21	If the youth is in school or foster care placement is in the youth's best interest; or if the placement is court ordered.
Montana	21	Youth may receive foster care maintenance payments beyond age 18 if the youth: 1) is at grade level, but will not graduate from high school until after turning age 18; 2) is at current grade level but is having academic difficulties or missing credits and will not graduate before turning age 18; or 3)is below grade level (one or more years behind his or her age group) and is more likely to graduate from high school if in care.
Nebraska	19	Age 19 is age of majority.
Nevada	21	*Did not respond to inquiry about conditions for remaining in care.*
New Hampshire	19	If youth has not graduated from high school before 19th birthday.
New Jersey	21	If the youth has not graduated from high school or needs special treatment that can not be provided through another source, or is a teen parent who needs to be under state supervision.
New Mexico	21	Youth do not stay in state legal custody after age 18. Youth may receive foster care maintenance payment if they receive independent living services.

Table B-1. (Continued).

State	Age at Which Youth May No Longer Remain in Foster Care	Conditions for Extended Foster Care (after 18th Birthday)
New York	21	*Did not respond to inquiry about conditions for remaining in care.*
North Carolina	21	If the youth signs a CARS agreement (Contractual Agreement for Residential Services) to remain in school or vocational training full time and live in a licensed foster care placement.
North Dakota	21	If the youth resides in a family foster home and is completing high school, or is attending an institution of higher education.
Ohio	21	If the youth is in school or has special needs.
Oklahoma	21	If youth has not finished high school or obtained a GED.
Oregon	21	If the youth is working on completing high school or obtaining a GED or has an Individualized Education Plan (IEP); or the case is reviewed by local child welfare agency and approved as an exception.
Pennsylvania	21	Before turning age 18, youth must ask court to retain jurisdiction to complete a course of treatment or education.
Puerto Rico		*Did not respond to survey.*
Rhode Island		*Did not respond to survey.*
South Carolina	21	If the youth signs agreement to remain in care while in school or vocational training; or due to a disabling condition that places youth at risk and remaining in care is in youth's best interest.
South Dakota	21	If the youth has not completed high school.
Tennessee		*Did not respond to survey.*
Texas	22	If the youth has not completed high school, GED program, or vocational training program.
Utah	19	With court order or significant need.
Vermont[a]	19	If the youth is enrolled in education program.
Virginia	21	If the youth is enrolled in an educational or vocational program and agrees to participate in the Independent Living Program.

State	Age at Which Youth May No Longer Remain in Foster Care	Conditions for Extended Foster Care (after 18th Birthday)
Washington	21	The youth must be attending high school or a vocational program.
West Virginia	21	If the youth is in school, has a Transitional Living Plan (TLP), or is employed. Youth may receive room and board until age 20 if: 1) youth requests continued financial supports after receiving a clear explanation of his or her right to independence and responsibility for self-support at age eighteen; and 2) youth will continue education (college, vocational, or training) and plans to continue current enrollment, or plans to enroll in a different school or training program within the next three months.
Wisconsin	19	If the youth is enrolled in high school or high school completion program.
Wyoming	21	Youth can remain in care until age 21 only for educational reasons. Court must approve the extended stay and case is reviewed very six months.
	Age 18 — 1 state; Age 19 — 6 states; Age 20 — 3 states; Age 21 — 32 states (including Washington, D.C.); Age 22 — 2 states; Age 23 — 2 states; No data provided — 6 states (including Puerto Rico)	

Source: Congressional Research Service presentation of data from University of Oklahoma, National Child Welfare Resource Center for Youth Development, 2006.

Note: This survey did not ask how many states encouraged youth to remain in care or how many youth actually remained in care.

[a] See Appendix C for information provided to the Congressional Research Service about foster care for youth ages 18 and older in these states.

APPENDIX C. DESCRIPTIONS OF FOSTER CARE FOR YOUTH AGES 18 TO 21 IN SELECT STATES

At least three states — Illinois, New York, and Vermont — as well as the District of Columbia provide foster care to young people ages 18 to 21 through state maintenance payments or similar types of payments. These payments are made out of state or local dollars and without regard to prior Title IV-E eligibility status. The Congressional Research Service contacted the four jurisdictions to learn more about their maintenance payment programs for older youth.[142] The jurisdictions provided varying levels of detail about their programs. *Note that this is not an exhaustive review of states that provide the payments to youth beyond their 18th birthdays.*

ILLINOIS

Illinois youth in foster care on their 18th birthday, regardless of their income or educational status, may stay in care (in a traditional foster care setting) until they are 19, and some continue in care until age 21. The majority of youth leave care at age 19. Until they age out of care, the Department of Child and Family Services (DCFS) DCFS continues to have guardianship of the youth and the courts retain jurisdiction of their cases. According to DCFS, Cook County, which includes the city of Chicago, tends to retain custody of youth for a longer period than other counties.

As of March 2008, about 1,520 youth ages 18 to 21 were in foster care homes. The state funds foster care for these older foster youth with state dollars and the average state foster care maintenance payment made to each foster household on their behalf is $458 per month. Youth not in foster family homes may be placed in an institution or group home, or other setting. The court continues to hold six-month reviews and annual review hearings for the youth.

Upon exiting foster care, youth remain under the guardianship of the state for six months, meaning that the court assigns DCFS the responsibility for the physical care and safety of the child, but does not make state foster care maintenance payments on their behalf. During and after the guardianship period, youth are eligible to receive transitional living services and housing through the state's independent living program. Youth attending college receive educational and other assistance from the state and remain in guardianship until age 21. This assistance is funded through state dollars, and not the federal Chafee Foster Care Independence program. As of March 2008, 815 youth ages 18 to 21 in guardianship lived in an independent living setting and 1,023 were in another setting, including college or a detention facility, or were on runaway status.

NEW YORK

New York enables youth to remain in foster care until age 21, provided the youth consents to remaining in care, if they are enrolled in a school, college, university, or vocational school, or if they lack the skills to live independently (generally due to cognitive and developmental deficiencies). As of December 2007, nearly 4,900 youth ages 17 through 20 were in foster care, of whom 1,994 were age 17; 1,269 were age 18; 924 were age 19; and 687 were age 20. The majority of the these youth had a permanency planning goal of discharge to another planned living arrangement (APPLA) with a permanency resource.[143]

The court continues to retain jurisdiction while the youth are in care (i.e., case hearings every six months and an annual permanency plan hearing) and the county provides the same case management services to youth (i.e., monthly case worker visits) as they received when they were under age 18. The county makes state foster care maintenance payments to foster parents for room and board, clothing, food, and other provisions, and youth continue to remain eligible for Medicaid. For youth who attend college and live on a college campus or in nearby housing, the local social services district pays the college directly for room and board.

Every youth who is discharged to APPLA with a permanency resource is first discharged on a trial basis, although youth can agree to be permanently discharged.[144] Trial discharge means the youth remains in the custody of the local social services district while living in the community. The same case planning and casework contact requirements apply as if the youth were still in a foster care placement, and permanency hearings are held. If a youth loses housing during the period of trial discharge, the district must assist the youth to find other appropriate housing or place the youth in a foster care setting. The trial discharge period may continue until a youth reaches the age of 21.

VERMONT

Youth in Vermont generally remain in foster care until age 18.[145] Effective July 1, 2007, Act 17 (2007) authorized funding for payments on behalf of former foster youth ages 18 through 22 under a program known as Extended Care, which includes the Adult Living program and the Housing Support and Incidental Grant program.[146] The Extended Care program is administered by the Vermont Department for Children and Families. Youth in the two programs are no longer in the custody of the state and the courts do not retain jurisdiction; youth voluntarily enroll in the program.[147] A youth development coordinator contracted with the Vermont Department of Children and Family Services works with youth in both programs (The Department contracts for independent living services in its 12 service districts.)[148]

The Vermont Legislature made a state FY2008 appropriation of $500,000 for the payments to caring adults (including family units) on behalf of older former foster youth. As of March 2008, the Extended Care program enrolled 42 youth. The budget for the program will likely support as many as 60 youth total.

Adult Living Program

The purpose of the Adult Living program is for youth to live with caring adults, known as adult partners, after they reach age 18 through the age of 22. Any eligible youth may decide at any point, up to age 21.5, to enroll in the Adult Living program. Participants in the program live in an arrangement similar to a foster family in households that are certified for a foster care placement, or a caring adult at least age 25 who undergoes a background check.[149]

To be eligible for the program, youth who have aged out of foster care must sign a voluntary services agreement pledging to gain independent living skills and work toward self sufficiency, and they must assume responsibility for expenses other than room and board. They are also to develop a savings plan in which they gradually save money until they become self sufficient, where possible. Youth must be enrolled in school part- or full-time, working, or actively seeking work. These criteria may be waived by the Department for Children and Families Services Commissioner or designee based on an individual youth's circumstance. Furthermore, if a youth is unemployed or not enrolled in an educational program, the department will determine if the youth is meeting the criteria for "productive time." Enrolled youth may live in college dormitories, but return to a foster care arrangement during school breaks. Youth who exit the program before age 21.5 may re-enter if they demonstrate a commitment to living within a home-like environment and working toward their goals of independent living.

For youth in the Adult Living program, the coordinator facilitates the youth working with their social worker and supportive adults to create and monitor a plan for life skills and self sufficiency; assists the youth in implementing the plan; calls and meets with him or her monthly; and submits monthly reports about the youth's status.

The reimbursement rate for the Adult Living program is $20.76 each day. Payments are made to adult partners for the cost of (and the cost of providing) basic provisions, including shelter, clothing, food, and school supplies. In the case of youth attending college, adult partners receive the payments to provide provisions to youth over their school breaks.

Housing Support Program

The Housing Support program provides financial support for youth ages 18 to 22 who live independently. Youth in the Housing Support program are also to sign the same voluntary services agreement that is signed by youth in the Adult Living program. The agreement includes information about how many hours they are to work and their budget. Youth may be asked to leave the program if they fail to fulfill the responsibilities outlined in the agreement; those who exit may return if they demonstrate a commitment to working toward their self-sufficiency goals. For youth in the Housing Support program, the coordinator assists the young person in preparing a housing contract; contacts the youth weekly in the first two months and monthly thereafter; assists with the young person's service coordination; and submits monthly reports about the status of the youth.

Youth in the Housing Support program receive a grant of up to $5,000 to cover housing and related expenses; on a case-by-case basis, youth may also receive supplemental funds for educational expenses, such as books and lab fees.[150]

WASHINGTON, D.C.

The Washington, D.C. Child and Family Service Agency (CFSA) requires that all foster youth remain in care until age 21 unless they (1) are reunified with their families or adopted; (2) get married; (3) join the military; or (4) can demonstrate that they have secured employment and housing and are self sufficient. Approximately 135 youth emancipate from the District of Columbia foster care system each year, of whom 95% are age 21; the remaining 5% meet the criteria listed above. Currently, the District has 446 youth ages 18 to 21 in care: 202 youth are age 18; 156 youth are age 19; and 117 youth are age 20.

Youth in care continue to remain under the jurisdiction of the court and receive the same case support from CFSA, except that social workers from the agency's Office of Youth Development manage their cases. Further, youth may live in a variety of settings — foster homes, group homes, and transitional living settings in an independent living setting, or in institutional settings. CFSA dollars from the D.C. government fund the foster care maintenance payments made to the foster parent or appropriate agency. Daily reimbursement rates range from $29.84 to $38.95, depending on whether the youth have special needs. These payments cover the cost of room and board, food, clothing, and a stipend, among other resources. The total cost of these services in FY2007, including maintenance payments, was $21.3 million (this includes youth in all types of living settings). Youth who are away at college receive housing payments and other assistance through D.C.'s independent living program.

APPENDIX D. FUNDING FOR THE CHAFEE FOSTER CARE INDEPENDENCE PROGRAM

Table D-1. Final FY2007 and Estimated FY2008 Federal CFCIP General and Voucher Allotments by State ($ in thousands)

State	Final FY2007 Allotments			Estimated FY2008 Allotments		
	General	Voucher	Total	General	Voucher	Total
Alabama	1,784	612	2,396	1,784	601	2,386
Alaska	500	159	659	500	156	656
Arizona	2,601	892	3,492	2,601	876	3,477
Arkansas	834	286	1,120	834	281	1,115
California	20,953	7,185	28,138	20,954	7,060	28,013
Colorado	2,120	727	2,847	2,120	714	2,834
Connecticut	1,815	622	2,438	1,815	612	2,427
Delaware	500	85	585	500	84	584
District of Columbia	1,092	222	1,314	1,092	218	1,310
Florida	7,566	2,595	10,161	7,566	2,549	10,116
Georgia	3,605	1,236	4,841	3,605	1,215	4,819
Hawaii	714	245	959	714	241	955
Idaho	500	161	661	500	158	658
Illinois	5,016	1,720	6,736	5,016	1,690	6,706
Indiana	2,906	996	3,902	2,906	979	3,885
Iowa	1,754	601	2,355	1,754	591	2,345
Kansas	1,506	516	2,023	1,506	507	2,014
Kentucky	1,881	645	2,526	1,881	634	2,515
Louisiana	1,358	428	1,786	1,358	420	1,778
Maine	596	204	800	596	201	797
Maryland	2,805	962	3,767	2,805	945	3,750

Table D-1. (Continued).

State	Final FY2007 Allotments			Estimated FY2008 Allotments		
	General	Voucher	Total	General	Voucher	Total
Massachusetts	3,161	1,084	4,244	3,161	1,065	4,225
Michigan	5,291	1,814	7,106	5,291	1,783	7,074
Minnesota	1,801	618	2,419	1,801	607	2,408
Mississippi	844	289	1,133	844	284	1,128
Missouri	2,928	1,004	3,932	2,928	987	3,915
Montana	574	197	770	574	193	767
Nebraska	1,608	552	2,160	1,608	542	2,150
Nevada	1,205	413	1,619	1,205	406	1,612
New Hampshire	500	104	604	500	102	602
New Jersey	3,108	1,066	4,174	3,108	1,047	4,156
New Mexico	591	203	794	591	199	791
New York	11,586	2,693	14,279	11,586	2,646	14,232
North Carolina	2,761	947	3,708	2,761	930	3,692
North Dakota	500	121	621	500	119	619
Ohio	4,502	1,544	6,046	4,502	1,517	6,019
Oklahoma	2,966	1,017	3,983	2,966	999	3,965
Oregon	2,845	976	3,820	2,845	958	3,803
Pennsylvania	5,599	1,920	7,519	5,599	1,887	7,486
Puerto Rico	1,752	601	3,503	1,752	590	2,342
Rhode Island	648	222	870	648	218	866
South Carolina	1,228	421	1,649	1,228	414	1,642
South Dakota	500	152	652	500	149	649
Tennessee	2,328	798	3,126	2,328	784	3,112
Texas	7,456	2,557	10,012	7,456	2,512	9,967
Utah	590	202	792	590	199	789
Vermont	500	127	627	500	125	625
Virginia	1,813	622	2,434	1,813	611	2,423
Washington	2,599	891	3,490	2,599	876	3,474
West Virginia	1,118	383	1,501	1,118	377	1,495
Wisconsin	2,093	718	2,810	2,093	705	2,798
Wyoming	500	112	612	500	110	610
State Subtotal	*137,900*	*45,464*	*184,515*	*137,900*	*44,671*	*182,571*
Technical Assistance	*2,051*	*0*	*2,051*	*2,100*	*0*	*2,100*
Set Asides	*0*	*645*	*645*	*0*	*680*	*680*
Total	140,000	46,110	186,061	140,000	45,351	185,351

Source: U.S. Department Health and Human Services, Administration for Children and Families, *FY2009 Justification of Estimates for Appropriations Committees, pp. D-89, D-90, G-24, G-25.*

Table D-2. Change in Funding by State from the Old Independent Living Program (as of FY1998) and the CFCIP (as of FY2007)

State	Funding in FY1998 ($)	Funding in FY2007 ($)	Percentage Change in Funding (%)
Alabama	1,038,490	1,784,444	71.8
Alaska	13,032	500,000	3736.7
Arizona	347,763	2,600,648	647.8
Arkansas	270,940	833,756	207.7
California	12,481,777	20,953,350	67.9
Colorado	825,854	2,120,011	156.7
Connecticut	754,518	1,815,162	140.6
Delaware	203,034	500,000	146.3
District of Columbia	1,091,992	1,091,992	0.0
Florida	987,045	7,566,271	666.6
Georgia	1,098,854	3,604,768	228.0
Hawaii	17,834	713,984	3,903.5
Idaho	107,004	500,000	367.3
Illinois	2,817,094	5,015,701	78.0
Indiana	1,019,970	2,905,756	184.9
Iowa	449,966	1,753,727	289.7
Kansas	717,477	1,506,181	109.9
Kentucky	791,557	1,880,984	137.6
Louisiana	1,358,131	1,358,131	0.0
Maine	565,888	596,019	5.3
Maryland	1,238,095	2,805,086	126.6
Massachusetts	635,852	3,160,529	397.1
Michigan	4,171,796	5,291,124	26.8
Minnesota	1,142,066	1,801,223	57.7
Mississippi	514,444	843,823	64.0
Missouri	1,295,026	2,928,213	126.1
Montana	244,190	573,562	134.9
Nebraska	435,562	1,608,401	269.3
Nevada	153,647	1,205,461	684.6
New Hampshire	320,326	500,000	56.1
New Jersey	2,297,848	3,108,387	35.3
New Mexico	207,149	591,373	185.5
New York	11,585,958	11,585,958	0.0
North Carolina	1,045,349	2,761,462	164.2
North Dakota	192,058	500,000	160.3
Ohio	2,860,992	4,502,283	57.4
Oklahoma	620,076	2,965,641	378.3
Oregon	930,799	2,844,837	205.6
Pennsylvania	4,638,225	5,599,072	20.7
Puerto Rico	N/A	1,751,663	N/A
Rhode Island	314,840	647,646	105.7

Table D-2. (Continued).

State	Funding in FY1998 ($)	Funding in FY2007 ($)	Percentage Change in Funding (%)
South Carolina	579,606	1,227,919	111.9
South Dakota	193,430	500,000	158.5
Tennessee	777,838	2,327,548	199.3
Texas	1,841,708	7,455,535	304.8
Utah	202,348	589,825	191.5
Vermont	295,633	500,000	69.1
Virginia	1,361,561	1,812,581	33.1
Washington	825,168	2,598,840	214.9
West Virginia	521,302	1,117,956	114.5
Wisconsin	1,554,305	2,093,167	34.7
Wyoming	44,585	500,000	1,021.5
State Subtotal	*70,000,000*[a]	*137,900,000*	*97.0*
Technical Assistance	*0*	*2,051*	*2,051.0*
Total	70,000,000	140,000,000	100.0

Source: Congressional Research Service presentation of data provided by the U.S. Department of Health and Human Services, Administration for Children and Families, May 2008.

N/A means not applicable.

Table D-3. FY2005 Chafee Foster Care Independence Program: Final Funds Allotted, Expended, and Returned to Federal Treasury, by State

State	Dollar Amount Allocated	Dollar Amount Expended	Dollar Amount Returned to the Treasury	Percent of Allotment Returned to Treasury
Alabama	1,563,344	1,563,344	0	0
Alaska	524,629	525,629	0	0
Arizona	1,991,020	1,991,020	0	0
Arkansas	771,514	771,514	0	0
California	25,012,729	25,012,729	0	0
Colorado	2,251,277	2,251,277	0	0
Connecticut	1,733,849	1,733,849	0	0
Delaware	500,000	499,958	42	0
District of Columbia	1,091,992	1,091,992	0	0
Florida	7,889,242	7,889,242	0	0
Georgia	3,506,787	3,506,787	0	0
Hawaii	763,027	763,027	0	0
Idaho	500,000	500,000	0	0

State	Dollar Amount Allocated	Dollar Amount Expended	Dollar Amount Returned to the Treasury	Percent of Allotment Returned to Treasury
Illinois	5,556,956	5,556,956	0	0
Indiana	2,288,567	2,248,212	40,355	1.8
Iowa	1,288,685	1,288,685	0	0.0
Kansas	1,486,707	1,486,707	0	0
Kentucky	1,773,196	1,334,896	438,300	24.7
Louisiana	1,358,131	1,358,131	0	0
Maine	771,257	771,257	0	0
Maryland	2,963,870	2,962,870	0	0
Massachusetts	3,242,415	3,128,310	114,105	3.5
Michigan	5,497,293	5,497,293	0	0
Minnesota	1,887,123	1,886,868	255	0
Mississippi	723,166	693,691	29,475	4.1
Missouri	3,090,942	2,638,275	452,667	14.6
Montana	500,000	500,000	0	0
Nebraska	1,553,057	1,553,057	0	0
Nevada	587,636	587,636	0	0
New Hampshire	500,000	500,000	0	0
New Jersey	3,298,993	3,298,993	0	0
New Mexico	540,060	540,060	0	0
New York	11,585,958	11,585,958	0	0
North Carolina	2,451,871	2,299,836	152,035	6.2
North Dakota	500,000	368,216	131,784	26.4
Ohio	4,969,320	4,969,320	0	0
Oklahoma	2,364,432	2,364,432	0	0
Oregon	2,412,523	2,412,523	0	0
Pennsylvania	5,598,104	5,598,104	0	0
Puerto Rico	1,950,644	1,950,644	0	0
Rhode Island	600,238	600,238	0	0
South Carolina	1,258,597	1,258,597	0	0
South Dakota	500,000	500,000	0	0
Tennessee	2,439,784	2,439,784	0	0
Texas	5,706,887	5,702,068	4,819	0
Utah	522,829	522,829	0	0
Vermont	500,000	500,000	0	0
Virginia	1,812,029	1,711,992	100,037	5.5
Washington	2,161,782	2,161,782	0	0
West Virginia	1,046,430	1,046,430	0	0
Wisconsin	2,012,108	2,012,108	0	0
Wyoming	500,000	497,614	2,386	0.5
Total	137,900,000[a]	136,433,740	1,466,260	1.1

Source: Congressional Research Service presentation of data from the U.S. Department of Health and Human Services, April 2008.

[a]. Training and technical assistance, research and evaluation, and set asides are not included in the total funding.

Table D-4. FY2005 Chafee Education and Training Vouchers: Estimated Funds Allotted, Expended, and Returned to Federal Treasury, by State

State	Dollar Amount Allocated	Dollar Amount Expended	Dollar Amount Returned to the Treasury	Percent of Allotment Returned to Treasury
Alabama	534,236	534,236	0	0
Alaska[a]	179,280	165,396	13,884	7.7
Arizona	680,385	680,385	0	0
Arkansas	263,647	226,720	36,927	14.0
California	8,547,517	8,451,971	95,546	1.1
Colorado	769,321	767,971	1,350	0.2
Connecticut	592,502	592,502	0	0
Delaware	71,536	71,536	0	0
District of Columbia	271,732	271,732	0	0
Florida	2,695,964	2,334,113	361,851	13.4
Georgia	1,198,362	1,198,362	0	0
Hawaii	260,747	260,747	0	0.3
Idaho	123,123	79,940	43,183	35.1
Illinois	1,898,960	1,898,960	0	0
Indiana	782,064	782,064	0	0
Iowa	440,378	440,378	0	0
Kansas	508,047	468,658	39,389	7.8
Kentucky	605,948	402,448	203,500	33.6
Louisiana	399,073	396,282	2,791	1.0
Maine	263,559	209,009	54,500	20.7
Maryland	1,012,491	700,931	311,560	30.8
Massachusetts	1,108,019	1,108,019	0	0
Michigan	1,878,571	1,128,776	749,795	39.9
Minnesota	644,880	595,124	49,756	7.7
Mississippi	247,125	186,275	60,850	24.6
Missouri	1,056,257	273,339	782,918	74.1
Montana	163,988	163,988	0	0
Nebraska	530,721	530,721	0	0
Nevada	200,811	200,811	0	0
New Hampshire	106,953	106,953	0	0
New Jersey[a]	1,127,354	1,127,354	0	0
New Mexico	184,553	78,401	106,152	57.5
New York	3,362,375	3,362,375	136,491	0

State	Dollar Amount Allocated	Dollar Amount Expended	Dollar Amount Returned to the Treasury	Percent of Allotment Returned to Treasury
North Carolina	837,869	825,205	12,664	1.5
North Dakota	108,798	77,470	31,328	28.8
Ohio	1,698,149	1,041,104	657,045	38.7
Oklahoma	807,989	807,989	0	0
Oregon	824,423	824,423	0	0
Pennsylvania	1,913,021	1,909,937	3,084	0.2
Puerto Rico[a]	666,587	18,209	648,378	97.3
Rhode Island	205,117	205,117	0	0
South Carolina	430,096	430,096	0	0
South Dakota	138,854	107,292	31,562	22.7
Tennessee	833,739	833,739	0	0
Texas	1,950,195	1,425,349	524,846	26.9
Utah	178,665	161,066	17,559	9.9
Vermont	123,826	123,826	0	0
Virginia	619,218	384,977	234,241	37.8
Washington	738,738	738,738	0	0
West Virginia	357,593	146,222	211,371	59.1
Wisconsin	687,591	641,074	46,517	6.7
Wyoming	92,716	92,716	0	0
Total [a]	45,923,663	40,591,026	5,332,637	11.6

Source: Congressional Research Service presentation of data from the U.S. Department of Health and Human Services, July 2007.

[a] Data are not yet final for Alaska, New Jersey, and Puerto Rico, and therefore the total data are not yet final.

Appendix E. Proposed Legislative Changes in the 110th Congress to the Chafee Foster Care Independence Program

Eight pending bills propose changes to the Chafee Foster Care Independence Program: Kinship Caregiver Support Act (H.R. 2188), introduced by Representative Danny Davis; Place to Call Home Act (H.R. 3409),[151] introduced by Representative Ruben Hinojosa; Tribal Foster Care and Adoption Access Act of 2007 (H.R. 4688 and S. 1956), introduced by Senator Max Baucus and Representative Earl Pomeroy; Reconnecting Youth to Prevent Homelessness Act (H.R. 4208 and S. 2560), introduced by Representative Shelley Berkeley and Senator John Kerry; and School Choice for Foster Kids Act (H.R. 4311), introduced by Representative Michele Bachmann; and the Improved Adoption Incentives and Relative Guardianship Support Act of 2008 (S. 3038), introduced by Senator Grassley. The changes include increased appropriations and expansion of the program, among other amendments.

Increased Appropriations

H.R. 3409 and H.R. 4208/S. 2560 propose to increase the annual appropriations for the CFCIP from $140 million to $200 million.

Eligibility

H.R. 2188, H.R. 3409, and H.R. 4208/S. 2560 would require states to provide educational, career, and other services to youth that states identify are likely to remain in foster care *until age 14*. Current law requires states to identify children likely to remain in care until age 18. H.R. 3409, H.R. 4208, and S. 2560 would further require states to provide aftercare services (housing, counseling, employment, education, and other appropriate services) to young adults *ages 18 to 25*. Currently, youth ages 18 to 21 are eligible for these services. The four bills would additionally require states to certify that they will provide support services, including housing, to youth as old as age 25. S. 3038 would add that a purpose of the CFCIP is to provide services and supports intended to help youth transition to independent adulthood to any youth ages 16 and older who have left foster care for relative guardianship or adoption.

Education and Training Vouchers

Currently, youth who were adopted from foster care at age 16 or older and youth who emancipate from care are eligible for the vouchers. H.R. 2188, H.R. 3409, and H.R. 4208/S. 5260 would require states to provide education and training vouchers to youth who left foster care at age 14 or older under a kinship guardianship arrangement or because they were adopted. The bills would also eliminate the upper age limit for eligibility purposes (currently 23) and the requirement that a youth must receive the voucher at age 21 to be eligible at age 22 or 23 for the program. S. 3038 would extend the vouchers to youth who left foster care at 16 years for relative guardianship or adoption.

Direct Funding for Tribal Entities

Under current law, tribes are not permitted to seek funding directly from the federal government for CFCIP funding.

Rather, they may enter into cooperative agreements with states for the tribe or the state to provide the independent living services to the youth.[152] S. 1956/H.R. 4688 would permit eligible tribes, tribal organizations, and tribal consortia to receive a part of a state's allotment of CFCIP and ETV funds (in proportion to the tribal entity's share of the foster care population living in the state) directly from the federal government in exchange for providing independent living services to tribal youth in the state who are aging out or are expected to age out of tribal care. Tribes would receive this funding whether they operate a Title IV-E foster care program (as provided under a separate provision of S. 1956/H.R. 4688); enter into a cooperative agreement with the state; or enter into an agreement with HHS (as provided under a separate provision of S. 1956/H.R. 4688). Alternatively, the bills would explicitly permit states and tribes to enter into cooperative agreements with the state to provide independent living services. The bills would require states to provide independent living services if the tribes decide not to provide the services under one of three pathways discussed above or enter into cooperative agreements with the state to provide the services.

PRIVATE SOCIAL SERVICE PROVIDERS

Current law does not specify the types of providers that are to assist the state in delivering support services to youth aging out of foster care; rather, it requires that states "involve the public and private sectors in helping adolescents in foster care achieve independence." H.R. 3409 and H.R. 4208/S. 2560 propose to amend the CFCIP by requiring states to distribute program funds to a range of qualified *private* social service providers and to ensure that these providers have equal opportunities to receive the funds.

INFORMATION ABOUT SUPPORT SERVICES

H.R. 3409 and H.R. 4208/S. 2560 would require that the state inform all children leaving care of the full range of available financial, housing, counseling, health, and other services for which the youth is eligible. The bills would further require that the HHS Secretary provide for the "efficient distribution" to states and local areas of information about federal programs — other than the CFCIP — that can assist youth in the transition to self-sufficiency and how to access the services under these programs.

EVALUATIONS

H.R. 3409 and H.R. 4208/S. 2560 would require the HHS Secretary to conduct evaluations of model independent living programs that focus on improving outcomes for youth aging out in a number of areas. The bills would also require the evaluations to include information on mental and physical health, personal development, and housing, as well as room and board services and how these services are improving housing outcomes for youth. The proposed provision would expand the evaluation requirements currently provided under law, which state that the HHS Secretary conduct evaluations of potential national significance and that the evaluations include information "on education, employment, and personal development."

REFERENCES

[1] U.S. Department of Health and Human, Administration for Children and Families, Children's Bureau. *The AFCARS Report #14: Preliminary FY2006 Estimates*, October 2006. At [http://www.acf.hhs.gov/programs/cb/stats_research/index.htm#afcars] (Hereafter referenced, U.S. Department of Health and Human Services, *AFCARS Report #14*.)

[2] Ibid.

[3] Fred Wulczyn, Lijun Chen, Kristen Brunner Hislop, *Foster Care Dynamics 2000-2005: A Report from the Multistate Foster Care Data Archive*, Chapin Hall Center for Children, University of Chicago, 2007.

[4] Bob Schoeni and Karen Ross, "Material Assistance Received from Families During the Transition to Adulthood." In Richard A. Settersten, Jr., Frank F. Furstenburg, Jr., and Rubén Rumbaut, eds., *On the Frontier of Adulthood: Theory, Research, and Public Policy*, pp. 404-405. Chicago: University of Chicago Press, 2005.

[5] According to the National Longitudinal Survey of Adolescent Health ("Add Health"), a nationally representative survey that tracks a cohort of youth over time, approximately 41% of surveyed youth age 21 in the study's third wave (2001-2002) lived with their parents. See Mark E. Courtney et al., *Midwest Evaluation of the Adult Functioning of Former Foster Youth: Outcomes at Age 21*, Chapin Hall Center for Children, University of Chicago, December 2007, p. 1. At [http://www.chapinhall.org/ article_abstract.aspx?ar=1355]. (Hereafter referenced Courtney et al., *Midwest Evaluation at Age 21*.)

[6] For a compendium of outcome studies and their methodologies, see Mark E. Courtney and Darcy Hughes Heuring. "The Transition to

Adulthood for Youth "Aging Out" of the Foster Care System" in Osgood et al., eds., *On Your Own Without a Net*, appendix. (Hereafter referenced as Mark E. Courtney and Darcy Hughes Heuring, The Transition to Adulthood for Youth "Aging Out" of the Foster Care System.")

[7] Fred Wulczyn et al. *Beyond Common Sense: Child Welfare, Child Well-Being, and the Evidence for Policy Reform* (New Brunswick: AldineTransaction, 2005), p. 116. (Hereafter referenced as Wulczyn et al., *Beyond Common Sense*.)

[8] Mark E. Courtney et al., *Midwest Evaluation of the Adult Functioning of Former Foster Youth: Outcomes at Age 19*, Chapin Hall Center for Children, University of Chicago, May 2005, pp. 68-70. At [http://www.chapinhall.org/ article_ abstract.aspx?ar=1355].

[9] Peter J. Pecora et al., *Improving Foster Family Care: Findings from the Northwest Foster Care Alumni Study*, Casey Family Programs, 2005. At [http://www.casey.org/Resources/ Publications/Northwest AlumniStudy.htm]. (Hereafter referenced as Peter J. Pecora et al., Improving Foster Family Care.)

[10] U.S. Department of Health and Human, *AFCARS Report #14*.

[11] These youth were placed in care prior to the enactment of the Chafee Foster Care Independence Act (P.L. 106-169) and most entered care because of sexual abuse and other type of maltreatment, which is not a primary reason for most children entering care (though the definition of sexual abuse in the U.S. Department of Health and Human Services' data collection system for children in foster care is not identical to the definition in this study). For additional information about the sample of youth, see pages 18 to 21 and 25 to 31 of the study.

[12] In a nationally representative study of children ages 11 to 14 entering foster care, 56.1% had a clinical/borderline score on the total problem behaviors checklist. Researchers often use this list as a proxy for mental health issues. Wulczyn et al., *Beyond Common Sense*, p. 108.

[13] Courtney et al., Midwest Evaluation at Age 21.

[14] Iowa amended its child welfare statute in 2006 to create a program that provides continuing support to foster youth ages 18 to 21. The Iowa youth in the Midwest study were already too old to benefit from the program when it was implemented.

[15] This information is based on an announcement made by Chapin Hall Center for Children, University of Chicago researchers at a December 9, 2007 Congressional briefing about the study's third wave of data.

[16] Mark E. Courtney, Sherry Terao, and Noel Bost, *Midwest Evaluation of the Adult Functioning of Former Foster Youth: Conditions of Youth Preparing to Leave State Care*, Chapin Hall Center for Children, University of Chicago, May 2005, pp. 28-30. At [http://www.chapinhall.org/article_abstract.aspx?ar=1355].

[17] Mark E. Courtney et al., *Midwest Evaluation of the Adult Functioning of Former Foster Youth: Outcomes at Age19*, Chapin Hall Center for Children, University of Chicago, May 2005, pp. 17-19. At [http://www.chapinhall.org/article_ abstract.aspx?ar= 1355].

[18] Courtney et al., *Midwest Evaluation at Age 21*, pp. 22-25.

[19] The Add Health sampled young adults from across the country, the majority (about 75%) of whom were white. Data from the Add Health survey were collected three to four years before wave 3 of the Midwest Evaluation. As a result, data on earnings and income have been adjusted by the Consumer Price Index (CPI) for comparison purposes.

[20] The written testimony from the hearing is available at [http://waysandmeans.house.gov/ hearings.asp?formmode= detail and hearing=612].

[21] Section 477(b)(3)(D).

[22] U.S. Department of Health and Human Services, Administration for Children and Families, Children's Bureau, *Child Welfare Policy Manual*, Section 3, Question 1. At [http://www.acf. hhs.gov/j2ee/programs/cb/laws_policies/laws/cwpm/policy.jsp?idFlag =3]. (Hereafter referenced as U.S. Department of Health and Human Services, Children's Bureau, *Child Welfare Policy Manual*.)

[23] Section 472(c) of the Social Security Act.

[24] U.S. Department of Health and Human Services, Administration for Children and Families, Children's Bureau, *AFCARS Report #14*.

[25] For additional information, see Section 8.3A, Question 2 of the *Child Welfare Policy Manual*.

[26] Jane Kim and Kevin Sobczyk, *Continuing Court Jurisdiction in Support of 18 to 21 Year-Old Foster Youth*, American Bar Association, Center on Children and the Law , July 2004, p. 16. At [http://www.abanet.org/child/court-jurisdiction.doc].

[27] See for example, Child Welfare League of America, "National News Roundup," Children's Voice, vol. 16, no. 9 (November/December 2007), p. 9; and "States Trying to Extend Foster Care Benefits," Stateline.org, August 23, 2007. At [http://www.stateline.org/ live/details/story?contentId=234381].

[28] U.S. Department of Health and Human Services, *AFCARS Report #14.*
[29] This information was provided to the Congressional Research Service by the state or city's child welfare or independent living services staff in March and April 2008.
[30] See Appendix E for proposed legislative changes to the CFCIP.
[31] U.S. Department of Health and Human Services, *Child Welfare Policy Manual*, Section 3.1G, Questions 1 and 4.
[32] Ibid, Section 3.1G, Questions 1 and 3.
[33] U.S. Department of Health and Human Services, *AFCARS Report #14.*
[34] U.S. Department of Health and Human Services, *Child Welfare Policy Manual*, Section 3.4, Question 5.
[35] National Foster Care Awareness Project, Frequently Asked Questions II: About the Foster Care Independence Act of 1999 and the John H. Chafee Foster Care Independence Program, December 2000, p16. At [http://www.casey.org/NR/rdonlyres/ E8E5EC9B-2C0B-496B-A165-5A55D2F793A5/459/ChafeeFAQI1.pdf]. (Hereafter referenced as National Foster Care Awareness Project, Frequently Asked Questions II.) Some of the members of the National Foster Care Awareness Project, which no longer exists, are now members of the National Foster Care Coalition. The National Foster Care Coalition produced the Frequently Asked Questions III publication about the program.
[36] U.S. Department of Health and Human Services, *Child Welfare Policy Manual*, Section 3.1F, Question 2.
[37] U.S. Congress, *Congressional Record*, July 1, 1999, p. S8124.
[38] U.S. Department of Health and Human Services, *Child Welfare Policy Manual*, Section 3.1B, Question 2.
[39] U.S. Department of Health and Human Services, *Child Welfare Policy Manual*, Section 3.1F, Question 3.
[40] National Foster Care Coalition, *Frequently Asked Questions III*, p. 9.
[41] U.S. Department of Health and Human Services, *Child Welfare Policy Manual*, Section 3.1I, Question 4.
[42] U.S. Congress, *Congressional Record*, June 25, 1999, p. H4969.
[43] Though most tribes provide child welfare services, 78 tribes and eight Alaska Regional Corporations have some form of a Title IV-E foster care agreement. This information is based on correspondence with the Association of American Indian Affairs, April 2008.
[44] U.S. Department of Health and Human Services, *Child Welfare Policy Manual*, Section 3.1I, Question 4.
[45] U.S. Department of Health and Human Services, *Child Welfare Policy Manual*, Section 3.1I, Question 5.

[46] University of Oklahoma, National Child Welfare Resource Center for Youth Development, *Tribal Youth Transitioning to Adulthood: Current Status of Independent Living Services Provided to Indian Youth*, November 2007. The NCWRCYD reviewed at least one of the following documents for each state: Child and Family Service Plans (CFSP) for FY2005-FY2009 and Annual Progress and Service Reports (ASPR) for FY2005 and FY2006.

[47] For additional information about the positive youth development movement in youth policy, see CRS Report RL33975, *Vulnerable Youth: Background and Policies*, by Adrienne L. Fernandes.

[48] National Foster Care Coalition, *Frequently Asked Questions II*, pp. 30-31.

[49] For a list of jurisdictions with youth advisory boards, see [http://www.fyi3.com/fyi3/ Involved/yabs/index.cfm].

[50] U.S. Department of Health and Human Services, *Child Welfare Policy Manual*, Section 3.1A, Question 1.

[51] For further information about the Jim Casey Youth Opportunities Initiative, see [http://www.jimcaseyyouth.org/communities.htm].

[52] U.S. Department of Health and Human Services, *Child Welfare Policy Manual*, Section 3.5C, Question 7.

[53] Section 472 of the Higher Education Act, as amended defines "cost of attendance" as tuition, fees, and other equipment or materials required of all students in the same course of study; books, supplies, and allowance for transportation and miscellaneous personal expenses, including computers; room and board; child care expenses for a student who is a parent; accommodations related to the student's disability that is not paid for by another source; expenses related to the youth's work experience in a cooperative education program; and student loan fees or insurance premiums on the loans. Section 102 of the Higher Education Act of 1965, as amended identifies "institutions of higher education" for purposes of student assistance (under Title IV of the Higher Education Act) to include traditional higher education institutions (i.e., public or private, nonprofit two- and four-year colleges and universities) as well as other postsecondary institutions (i.e., proprietary or for-profit schools offering technical training programs usually of less than two-years' duration, and vocational schools).

[54] National Foster Care Coalition, *Frequently Asked Questions III*, p. 12.

[55] U.S. Department of Health and Human Services, *Child Welfare Policy Manual*, Section 3.5C, Question 5.

[56] National Foster Care Coalition, *Frequently Asked Questions III*, p. 7.
[57] Ibid, Section 3.5C, Question 5.
[58] Ibid, Section 3.5C, Question 6.
[59] For example, see National Governors Association, Center for Best Practices, *State Policies to Help Youth Transition Out of Foster Care*, Issue Brief, December 2007. (Hereafter referenced as National Governors Association, *State Policies to Help Youth Transition Out of Foster Care*.)
[60] U.S. Department of Health and Human Services, *Child Welfare Policy Manua*l, Section 3.5, Question 1.
[61] National Foster Care Coalition, *Frequently Asked Questions III*, p. 7.
[62] U.S. Department of Health and Human Services, Administration for Children and Families, *FY2009 Justification of Estimates for Appropriations Committees,* p. D-87.
[63] This information is based on CRS analysis of FY1998 and FY2007 funding data for each state. This analysis includes only the 50 states and the District of Columbia.
[64] This information was provided to CRS in June 2007.
[65] U.S. Department of Health and Human Services, "Chafee National Youth in Transition Database," 73 *Federal Register* 10338, February 26, 2008.
[66] For additional information, you may request a copy of a Congressional Distribution Memorandum, *Chafee Foster Care Independence Act National Youth in Transition Database*, by Adrienne L. Fernandes.
[67] Additional information regarding the Multi-Site Evaluation of Foster Youth Programs is available at [http://www.acf.hhs.gov/programs/opre/project/tprojectIndex.jsp?topicId=2].
[68] This information was provided to the Congressional Research Service by U.S. Department of Health and Human Services, Administration for Children and Families, Office of Planning, Research and Evaluation on February 11, 2008.
[69] U.S. Office of Management and Budget, *Detailed Information on the Independent Living Program Assessment*, 2004. At [http://www.whitehouse.gov/omb/expectmore/summary/10002146.2004.html].
[70] Other resources illustrate how select jurisdictions and programs provide independent living services for older current and former foster youth. See, for example, U.S. Government Accountability Office, *Disconnected Youth: Federal Action Could Address Some of the Challenges Faced by Local Programs That Reconnect Youth to*

Education and Employment, GAO-08-313, February 2008; National Governors Association, Center for Best Practices, State Policies to Help Youth Transition Out of Foster Care, Issue Brief, January 2007; Wilhelmina A. Leigh et al., Aging Out of the Foster Care System to Adulthood: Findings, Challenges, and Recommendations, Joint Center for Political and Economic Studies and Black Administrators in Child Welfare Inc., December 2007; and Rachel H. Sherman, "Serving Youth Aging Out of Foster Care," Welfare Information Network Issue Note, vol. 8, no. 5 (October 2004), pp. 5-7. See also the University of Chicago Law School, "Transition From Foster Care to Adulthood Wiki" at [http://fostercaretoadulthood.wiki spaces.com/] and the University of Oklahoma, National Child Welfare Resource Center for Youth Development, "State by State Facts Page," at [http://www.nrcys.ou.edu/yd/state_ pages.html].

[71] For additional information, see CRS Report RL32968, *Child Welfare: State Performance on Child and Family Services Reviews*, by Emilie Stoltzfus.

[72] In 2000, HHS published a final rule to establish a formal review process consistent with legislative mandates to improve federal oversight of state child welfare programs. The final rule established the CFSR and Title IV-E Foster Care Eligibility Reviews. The Foster Care Eligibility Reviews are conducted to validate a state's claim for federal reimbursement of payments made on behalf of eligible children, and are not discussed in this report.

[73] The reports are available at [http://basis.caliber.com/cwig/ws/ cwmd/docs/cb_web/ SearchForm].

[74] One item used in reviews conducted in 2001 assessed state provision of independent living services for children age 16 or older (Item 8). However, for reviews in 2002 to 2004, review teams looked instead at appropriate and timely achievement of reunification, guardianship, or kinship placement.

[75] U.S. Department of Health and Human Services, Administration for Children and Families, Children's Bureau, *Child and Family Services Reviews Procedures Manual, Working Draft*, November 2006.

[76] Edi Winkle, Dorothy Ansell, and Ann Newman, *An Analysis of State's Child and Family Services Reviews and Program Improvement Plans From a Youth Development Perspective*, National Resource Center for Youth Development, University of Oklahoma, March 25, 2004, at [http://www.nrcys.ou.edu/yd/ resources/ publications/ pdfs/

summaryv2.2.pdf]. This analysis was based on 45 of the final reports that were available as of March 15, 2004.

[77] Ibid.

[78] U.S. Government Accountability Office. HHS Actions Could Improve Coordination of Services and Monitoring of States' Independent Living Programs, GAO-05-25, November 2004, pp. 18-19.

[79] National Foster Care Coalition, *The Chafee Education and Training Voucher Program: Six States' Experiences*, National Foster Care Coalition and Casey Family Programs, 2007. At [http://www.casey.org/Resources/Publications/ETV.htm].

[80] In a separate publication, from 2005, the National Foster Care Coalition identified promising practices for ETV programs. These include (1) a clearly defined application process and a funding process transparent to stakeholders, including informing students of the amount of funding they receive; (2) open lines of communication between applicants/participants and ETV administrators, including multiple methods for contacting the ETV office; and (3) individual assessments for all applicants to ensure that ETV funds are based on their unmet financial need as calculated by their educational institution; among other approaches. See National Foster Care Coalition, *Frequently Asked Questions III: About the Chafee Foster Care Independence Program and the Chafee Educational and Training Program*.

[81] Section 1902(a)(10)(A)(ii)(XVII) of the Social Security Act.

[82] Section 1905(w)(1)) of the Social Security Act.

[83] Sonali Patel and Martha A. Roherty, *Medicaid Access for Youth Aging Out of Foster Care,* American Public Human Services Association, 2007. At [http://www.aphsa.org/ Home/Doc/ Medicaid-Access-for-Youth-Aging-Out-of-Foster-Care-Rpt.pdf]. Puerto Rico was not included in this analysis. (Hereafter referenced as Sonali Patel and Martha A. Roherty, *Medicaid Access for Youth Aging Out of Foster Care*.) For additional informmation about Missouri's coverage option, see [http://www.senate.mo.gov/07info/BTS_Web/Bill. aspx? Session Type=R and BillID=28834].

[84] Abigail English, Amy J. Stinnett, and Elisha Dunn-Georgiou, *Health Care for Adolescents and Young Adults Leaving Foster Care: Policy Options for Improving Access*, Center for Adolescent Health and the Law, p. 5, February 2006, at [http://www.cahl.org/PDFs/FC IssueBrief.pdf].

[85] For information about Medicaid eligibility pathways, see CRS Report RL33019, *Medicaid Eligibility for Children and Adults*, by Jean Hearne.

[86] Though not discussed here, a small part of the allocation formula population factor for the Title I-A program of Education for the Disadvantaged (authorized under the Elementary and Secondary Education Act, as amended) accounts for the number of children ages five to 17 who are in institutions for delinquent children or foster homes when making grants to local education agencies (LEAs). For additional information, see CRS Report RL33731, Education for the Disadvantaged: Reauthorization Issues for ESEA Title I-A Under the No Child Left Behind Act, by Wayne C. Riddle.

[87] The previous definition included an individual who is an orphan or ward of the state (or was until age 18). 29 U.S.C. 1087vv(d). The College Cost Reduction and Access Act Technical Amendments of 2007 (H.R. 4153), would make a correction to the definition of independent student to include any child who "is an orphan, in foster care, or a ward of the court, or was an orphan, in foster care, or a ward of the court any time when the individual was 13 years of age or older."

[88] Authorization of appropriations under WIA expired in FY2003 but is annually extended through appropriations acts. Youth in foster care are also eligible for WIA's Youth Opportunity program, however, Congress has not appropriated funding for the program since FY2003.

[89] 29 U.S.C. 2801(13) and 29 U.S.C. 2884(1).

[90] 29 U.S.C. 2916.

[91] Institute for Educational Leadership, Foster Care Youth Employment Demonstration Project: Summary Report of Year Two Site Visit, May 2007, at [http://www.iel.org/pubs/ caseysitesreport_year2.pdf].

[92] For additional information, see [http://www.iel.org/ programs/ casey.html].

[93] Institute for Educational Leadership, *Foster Care Youth Demonstration Project: Draft Final Evaluation Report, Executive Summary*, forthcoming in June 2008. The report will be available at [http://www.casey.org] and [http://www.iel.org].

[94] 42 U.S.C. 1437(f)(x).

[95] U.S. Congress, House Committee on Appropriations, *Joint Explanatory Statement, Division K.* report to accompany FY2008 Consolidated Appropriations Amendment to H.R. 2764 (P.L. 110-

161), 110th Cong., 1st sess.,p. 2396. At [http://www.gpoaccess.gov/congress/ house/appropriations/08conappro.html].

[96] For additional information, see CRS Report RL33785, *Runaway and Homeless Youth: Demographics, Programs, and Emerging Issues*, by Adrienne L. Fernandes.

[97] Data on youth served by the program are provided in HHS's National Extranet Optimized Runaway and Homeless Youth Management Information System (NEO-RHYMIS) Available at [https://extranet.acf.hhs.gov/rhymis/custom_reports.html.]

[98] National Alliance to End Homelessness, "Federal Funding for Youth Housing Programs," information presented at National Alliance to End Audio Conference, March 9, 2006. Available at [http://naeh.org/content/article/browse/?type=24andtopic=Youth].

[99] U.S. Department of Health and Human Services, "Training of Child Welfare Agency Supervisors in the Effective Delivery and Management of Federal Independent Living Service for Youth in Foster Care," 70 *Federal Register* 35087, June 16, 2005.

[100] In the fall of 2000, HHS awarded twelve grants for Independent Living Training for Child Welfare practitioners. One of the findings from the completed projects was that child welfare supervisors needed training on youth development to understand the unique developmental and service needs of youth in care.

[101] For an overview of current child welfare legislation, including bills that pertain to older youth in care or youth emancipating from foster care, see CRS Report RL34388, *Child Welfare Issues in the 110th Congress*, by Emilie Stoltzfus.

[102] Written testimony is available at [http://waysandmeans.house.gov/hearings.asp?form mode=detail and hearing=576].

[103] Iowa amended its child welfare statute in 2006 to create a program that provides continuing support to foster youth ages 18 to 21. The Iowa youth in the Midwest study were already too old to benefit from the program when it was implemented.

[104] Mark E. Courtney, Amy Dworsky, and Harold Pollack, *When Should the State Cease Parenting? Evidence from the Midwest Study*, Chapin Hall Center for Children, University of Chicago, Issue Brief no. 115, December 2007. At [http://www. chapinhall. org/article_ abstract.aspx?ar=1355], visited March 24, 2008.

[105] The statute provides that eligible group facilities may be publicly or privately operated, but in the case of publicly operated facilities, they may not have more than 25 beds.

[106] See U.S. Department of Health and Human Services, Administration for Children and Families, *Child Welfare Policy Manual,* Section 8.3B, Question 2, regarding allowable use of Title IV-E foster care maintenance payments.

[107] Mark E. Courtney et al., *Midwest Evaluation of the Adult Functioning of Former Foster Youth: Outcomes at Age 19,* Chapin Hall, University of Chicago, May 2005, p. 13. At [http://www.chapinhall.org/article_abstract.aspx?ar=1355].

[108] S. 1512 would make "conforming amendments" to some of the purposes of the Chafee Foster Care Independence (CFCIP) program. Overall, these amendments would change all references to youth who aged out, or who are expected to age out of foster care at age 18, to reference instead any older age that the state chooses as the age at which all youth must leave foster care custody (i.e., at the 19th, 20th or 21st birthday). These "conforming" changes to the purposes of CFCIP would not appear to prohibit services to currently eligible individuals, however, by referencing an older age in the context of identifying and providing independent living services, they might suggest a later onset of services. The same kind of "conforming" amendments would also be made to certain stipulations that states must abide by in order to receive CFCIP funds. For additional information about these and other changes made by S. 1512, see CRS Congressional Distribution Memorandum *Foster Care Continuing Opportunities Act (S. 1512),* by Emilie Stoltzfus and Adrienne Fernandes.

[109] For additional information about the need for permanency for current and former older foster youth, see Benjamin Kerman and Madelyn Freundlich, *Recommendations for Policy, Practice and Research, Proceedings from the Research Roundtable at the 2006 National Convening on Youth Permanence*, Annie E. Casey Foundation and Excal Consulting Partners, 2006,at [http://www.casey familyservices.org/pdfs/permanency_proceedings_final .pdf]; and Gina Miranda Samuels, *A Reason, A Season, or A Lifetime: Relational Permanence Among Young Adults with Foster Care Backgrounds*, Chapin Hall Center for Children, University of Chicago, 2008, at [http://www.chapinhall.org/ article_abstract.aspx? ar=1466 and L2=61 and L3=130].

[110] Written testimony is available at [http://waysandmeans.house.gov/hearings.asp?form mode=detail and hearing=576].

[111] National Governors Association, State Policies to Help Youth Transition Out of Foster Care, pp. 4-7.

[112] This information was provided by Alameda County Children and Family Services in May of 2007 at the National Pathways to Adulthood Conference.

[113] The University of Oklahoma, National Child Welfare Resource Center for Youth Development, *The John H. Chafee Foster Care Independence Program: Aftercare Services*, 2003. (Hereafter referenced as The University of Oklahoma, National Resource Center for Youth Development, Aftercare Services). At [http://www.nrcys.ou.edu/yd/resources/publica tions/monographs/aftercare.pdf].

[114] Peter J. Pecora et al., *Improving Foster Family Care*, p. 41.

[115] Mark E. Courtney et. al., Midwest Evaluation of the Adult Functioning of Former Foster Youth: Outcomes at Age 21, pp. 15-16.

[116] Miryam J. Choca et al. "Can't Do It Alone: Housing Collaborations to Improve Foster Youth Outcomes." In Ruth Anne White and Debra J. Rog, eds., *Child Welfare*, vol. LXXXIII, no. 5 (September/October 2004), pp. 469-474.

[117] Mark E. Courtney and Darcy Hughes Huering, "The Transition to Adulthood for Youth "Aging Out" of the Foster Care System," p. 54.

[118] The University of Oklahoma, National Child Welfare Resource Center for Youth Development, *Aftercare Services*, pp. 24-26.

[119] Written testimony is available at [http://waysandmeans.house.gov/hearings.asp?form mode=detail and hearing=576].

[120] For additional information about this population, see CRS Report RL33785, *Runaway and Homeless Youth: Demographics, Programs, and Emerging Issues*, by Adrienne L. Fernandes.

[121] Caren Kapplan, *Children Missing from Care*, Child Welfare League of America, 2004, at [http://www.cwla.org/ programs/ fostercare/ childmiss.htm].

[122] U.S. Department of Health and Human Services, *AFCARS Report #14*.

[123] Mark E. Courtney et al., *Youth Who Run Away from Substitute Care*, Chapin Hall Center for Children, University of Chicago, Issue Brief no. 103, March 2005, p. 2, at [http://www.chapinhall.org/article_abstract.aspx?ar=1382].

[124] This provision applies to all children and is defined in detail at Section 475(5) of the Social Security Act.

[125] Pew Charitable Trusts and Jim Casey Youth Opportunities Initiative, *Time for Reform: Aging Out and On Their Own*, 2007, p. 12, at [http://kidsarewaiting.org/publications/ reports].

[126] See Mark E. Courtney et. al., Midwest Evaluation of the Adult Functioning of Former Foster Youth: Outcomes at Age 21 and Peter J. Pecora et al., Improving Foster Family Care.

[127] Under Title IV-E, states may not make foster care maintenance claims for the cost of transporting children to their school of origin because education is not included in the definition of a foster care maintenance payment. However, this type of cost can be an administrative cost because it is a part of case management. To make an administrative claim for school transportation costs, a state must discuss this kind of claim in its approved cost allocation plan. U.S. Department of Health and Human Services, Administration for Children and Families, Children's Bureau, *Child Welfare Policy Manual*, Section 8.1B, Question 27.

[128] Written testimony is available at [http://waysandmeans.house.gov/hearings.asp?formmod e=detail and hearing=569].

[129] Mark E. Courtney et al., *Midwest Evaluation of the Adult Functioning of Former Foster Youth: Outcomes at Age 19,* p. 44. Although Iowa is one of 17 states that have taken up the Chafee Medicaid option, some former foster youth appear not to be covered, perhaps because they do not meet the eligibility criteria for coverage.

[130] Ibid, pp. 39-40.

[131] To be eligible for Medicaid coverage in Iowa from age 18 to age 21, youth must have exited foster care at age 18 and have countable income under 200% of the federal poverty line.

[132] APHSA is a non-profit trade organization of state and local human service agencies and individuals who work in or are interested in public human service programs. See Sonali Patel and Martha A. Roherty, *Medicaid Access for Youth Aging Out of Foster Care*, pp. 3-6.

[133] Section 1902(a)(10)(A)(i)(I) of the Social Security Act.

[134] Federal law defines over 50 distinct population groups as being potentially eligible for states' Medicaid programs. Some groups are mandatory, meaning that federal law requires all states and the District of Columbia that participate in Medicaid to cover them. Other groups are optional; that is, federal law allows states to choose to cover them. For additional information on Medicaid eligibility for adults and children, see CRS Report RL33019, *Medicaid Eligibility for Adults and Children*, by Jean Hearne.

[135] U.S. Government Accountability Office, Disconnected Youth: Federal Action Could Address Some of the Challenges Faced by Local

Programs That Reconnect Youth to Education and Employment, GAO-08-313, February 2008.

[136] The written testimony from the hearing is available at [http://waysandmeans.house.gov/ hearings.asp?formmode=detail and hearing=569].

[137] For more details about the analysis of disconnected youth, see CRS Report RL33975, Vulnerable Youth: Background and Policies, by Adrienne L. Fernandes.

[138] For additional information, see CRS Report RL31622, *Trio and GEAR UP Programs: Status and Issues*, by Jeffrey J. Kuenzi.

[139] H.R. 4137 defines current and former foster youth as "orphans, in foster care, or wards of the court, or who were in foster care or were wards of the court until the students reached the age of 16," as part of the term "disconnected students."

[140] For additional information, see CRS Report RS22185, Individual Development Accounts (IDAs): Background and Current Legislation for Federal Grant Programs to Help Low-Income Families Save, by Gene Falk.

[141] This information was provided to the Congressional Research Service by the Jim Casey Youth Opportunities Initiative in February 2008.

[142] This information was provided to the Congressional Research Service by the state or city's child welfare or independent living services staff in March and April 2008.

[143] A permanency resource is a caring adult willing to help provide emotional support and guidance to a youth as a youth transitions to adulthood, and can be a foster parent, teacher, parent of a friend in the community.

[144] Youth in care ages 16 and 17 must be discharged on a trial basis if they have been in care for 12 of the last 36 months and if their case goal is "APPLA with a permanency resource." These youth may decline the trial discharge and opt instead for a final discharge; they may also return to care if they are discharged.

[145] As of March 2008, the state was using an interim plan to provide guidance about the program. The state expects to promulgate regulations establishing the program guidelines by August 1, 2008.

[146] A third component of the program, not discussed here, offers supports and services on behalf of youth who reach their 18th birthday while in custody and are attending high school.

[147] The legislature also appropriated $203,000 to fund 2.5 full-time youth development coordinator staff to serve the older foster (and

emancipated) youth population, as well as to support driver's education, prevention of homelessness, and transportation to the youth's school of origin.

[148] All youth who emancipate from foster care are eligible to work with the youth development coordinators and for other supports, such as housing and workforce training, through the state's independent living program.

[149] The Department for Children and Families plans to address the feasibility of this second option in its regulations.

[150] Youth may receive direct payment under the Housing Support and Incidental Living Grants program. Yet these youth might be rendered ineligible for Temporary Assistance for Needy Families (TANF) because the payment would count as direct income for TANF eligibility purposes. If, however, the payment is made directly to a youth's vendor (i.e., landlord or college) this assistance is considered a "subsidy."

[151] H.R. 3409 is an omnibus child welfare and youth policy bill that makes changes to several federal programs concerning youth, primarily those who have run away or are homeless.

[152] This arrangement is not explicit under current law.

INDEX

C

D

E

F

G

H

I

Q

R

S

T

U

V

W

Y